T0258862

Future Trends and Challenges for ICT Standardization

RIVER PUBLISHERS SERIES IN STANDARDISATION

Volume 3

Standardisation is a book series addressing the pre-development related standards issues and standardized technologies already deployed. The focus of this series is also to examine the application domains of standardised technologies. This series will present works of foras and standardization bodies like IETF, 3GPP, IEEE, ARIB, TTA, CCSA, WiMAX, Bluetooth, ZigBee, etc.

Other than standards, this book series also presents technologies and concepts that have prevailed as *de-facto*.

Scope of this series also addresses prevailing applications which lead to regulatory and policy issues. This may also lead towards harmonization and standardization of activities across industries.

For a list of other books in this series, see final page.

Future Trends and Challenges for ICT Standardization

Editor

Ramjee Prasad

Founder Chairman, GISFI and
Director CTIF, Aalborg University, Denmark

River Publishers

Aalborg

ISBN 978-87-92329-38-7 (paperback)

Published, sold and distributed by:
River Publishers
P.O. Box 1657
Algade 42
9000 Aalborg
Denmark

Tel.: +45369953197
www.riverpublishers.com

All rights reserved © 2010 River Publishers

No part of this work may be reproduced, stored in a retrieval system, or trans-mitted in any form or by any means, electronic, mechanical, photocopying, microfilming, recording or otherwise, without prior written permission from the Publisher.

To the ICT Industry, Academia and Government of India
for jointly stepping towards the standardization of the
new era in technological and economic evolution of India

Contents

Preface

इन्द्रियाणि पराण्याहुरिन्द्रियेभ्यः परं मनः ।
मनसस्तु परा बुद्धिर्यों बुद्धेः परतस्तु सः ॥

indriyani parany ahur indriyebhyah param manah
manasas tu para buddhir yo buddheh paratas tu sah

The senses are superior to other parts of the body. The mind is superior to the senses. The intellect is superior and stronger than the mind and superior and stronger than the intellect is the soul.

The Bhagvad-Gita (3.42)

The Global ICT Standardization Forum for India (GISFI) is an Indian standardization body active in the area of Information and Communication Technologies (ICT). During my meeting together with Korean, Japanese and Chinese companies and other organizations of the various countries, I conceived the idea of a standardization body for India, which would represent the nation in totality. Soon after realizing the imperativeness of such an entity for India's growth in the knowledge based economy, I proposed the matter to several major Indian companies and they agreed to the immediate need for such an initiative. Furthermore, the international companies and organizations also expressed that they face difficulties in entering the Indian market. The underlying issue is fragmentation of the market, and due to this disparateness, the country has lost the opportunity for technological growth.

Global ICT Standardization Forum for India (GISFI) was inaugurated on 29–30 May 2009.

GISFI has been set up as a public-private partnership, which will include industry, academia, manufacturers, operators, standardization bodies, and the

Government. The main objective of GISFI will be to harmonise and unify the standardization efforts within the Indian market and work in close coopera-tion with all standardization bodies acting as a catalyst for India's growth as a knowledge-based economy. This will also result in improved competitive-ness within companies and create a conducive investor-friendly environment. GISFI will provide a platform for discussion where issues related to stand-ardization will be deliberated, information exchanged, and knowledge shared between all stakeholders.

GISFI is an effort to create a new coherence and strengthen the role of India in the world standardization process by mapping the achievements in ICT in India to the global standardization trends. Further, GISFI is focused on strengthening the ties among leading and emerging scholars and institutions in India and the world to develop and cultivate a research and development agenda for the field.

GISFI addresses the research and product development of ICT in India and provides a bridge towards the globalization of the Indian achievements. The issues of technology, governance, and development and a platform for raising an awareness of the importance and the internationalization of higher education in the field are supported by the joint partnership with the Gov-ernment of India. The working groups organized in GISFI will draw know-ledge from academia, business, civil society, and Government/policy-making circles. A strong global platform is the key to a successful standardization process that can benefit all user groups.

In a nutshell, GISFI encompasses activities such as drawing up provi-sions and requirements for the ICT standardization process, using existing solutions (set down in standards), and bringing India's own procedure into line so that cooperation is simpler and purchasing cheaper.

During the discussions and deliberations at the GISFI workshops, it has been realized that a book should be published in the areas in which GISFI is pursuing the standards. The Information technologies communicate the information between two devices, or human beings according to users' re-quirements by means of wired or wireless IP networks to give the best Quality of Services . Indirectly all technologies related with communication are re-quired to be integrated to deliver the best economic and generic product for users' minute requirements. This book is the outcome of the integration of technologies in the areas such as RF Spectrum management, Future radio access, Convergence, Internet of Things and Green ICT.

Overview of the Book

This book aims at identifying the importance of ICT standardization for strengthening the Indian industrial and business sector. Further, it outlines the major challenges and trends in the ICT development worldwide, while mapping the Indian effort on the background of the overall progress.

This book is organized as follows: Chapter 1 outlines the general sequence of the book, providing a rapid overview of future technologies in the areas of spectrum management, future radio access technologies, Convergence, Internet of Things and Green ICT.

Chapter 2 describes how new spectrum technologies and management methods might thoroughly change the way we look at the future spectrum usage with spectrum trading and clearing (re-allocation), software defined radio: dynamic (versus static) allocation of the spectrum, mesh networks, frequency sharing and hopping, frequency on demand, optimization of non-permanent usages, increased system (emitter and receiver) resilience and tolerance to interference. It outlines the importance of spectrum for the development of ICT and the challenges that standardization is facing.

Chapter 3 identifies the requirements and challenges for future radio access technologies which are discussed with respect to data rate, spectrum, energy efficiency and advanced channel models. The need for improving the spectrum efficiency is discussed along with multiple antenna systems, relaying and base station cooperation, and heterogeneous network environments. SDR and cognitive radio are also discussed as enablers of novel radio access.

Chapter 4 describes the issues related to the convergence of services like voice, video and data in the next generation networks and how it can lead to more competition in individual markets for each of these services. Convergence of services puts an increasing competitive pressure on mobile carriers from the IP world. However, the trend towards horizontal integration of infrastructures, market and services could lead to strengthening of market power as there may be relatively few companies that can package voice, video and data services in a single bundled offer to end users.

Chapter 5 identifies the challenges and outlines solutions for the realization of the real potential of the ICT area of IoT. IoT can provide services that can be used by millions of independent devices co-operating together in large or small combinations, and in shared or separated federations. This requires a common a networking platform and a set of commonly agreed methods of information exchange and operation.

Chapter 6 is about environmental sustainability through use of ICT and how current ICT can be optimized towards more energy-efficiency. Climate change is now the greatest challenge in the global community is faced with – having serious implications on the future of world economy, health and safety, food, security and many other dimensions. Green ICT is described as a set of evolving tools, methods, techniques and applications that can be used for reducing the carbon footprint of all human activity including the ICT sector itself.

Finally, Chapter 7 concludes the book.

Completing the first book of the GISFI gives me immense pleasure. This book will help in setting up the standards for India. Any remarks to improve the text and to correct the errors would be highly appreciated.

Ramjee Prasad

Acknowledgements

I would like to express my heartfelt gratitude to colleagues from the Government, industries and academia without whom, this book would have never been completed. On behalf of GISFI, I would like to appreciate the effort put in by: Albena Mihovska, Debassis Bandyopadhyay, Jaydip Sen, Rajeev Prasad, Parag Pruthi, U.B. Desai, D.K. Singh, 11 GISFI PhDs (Kishor Puna Patil, Dipashree Bhalerao, Borhade Ravindra Honaji, Nandkumar P. Kulkarni, Sachin Dilip Babar, Madhukar M. Deshmukh, Sulakshana Patil, Manohar Chaudhari, Parikshit Narendra Mahalle, Shivprasad P. Patil, Vandana Milind Rohokale), and the contributors of all chapters.

Finally, special thanks go to Jaydip Sen for the overall coordination in editing the book.

List of Figures

List of Tables

List of Acronyms

3GPP	3rd Generation Partnership Project
AAA	Authentication Authorization and Accounting
ADC	Access Deficit Charge
AES	Advanced Encryption Standard
AmI	Ambient Intelligence
API	Application Programming Interface
ARIB	Association of Radio Industries and Businesses
ARPU	Average Revenue Per User
AS	Application Server
ATIS	Alliance for Telecommunication Industry Solutions
BAN	Body Area Networks
BBC	British Broadcasting Corporation
BDA	Broadband Development Authority
BGCF	Breakout Gateway Control Function
BPL	Broadband over Power Line
BSNL	Bharat Sanchar Nigam Limited
BWA	Broadband Wireless Access
CAGR	Compound Annual Growth Rate
CAMEL	Customized Applications for Mobile Network Enhanced Logic
CAP	Camel Application Part
CAPEX	Capital Expense
CASAGRAS	Coordination and Support Action for Global RFID-Related Activities and Standardisation
CATV	Cable Television
CBC-MAC	Cipher Block Chaining Message Authentication Code
CCK	Complementary Code Keying
CDMA	Code Division Multiple Access
CDR	Charging Data Record
CN	Core Network
CooP MAC	Cooperative Multiple Access

COPS	Common Open Policy Service
COPS-PR	Common Open Policy Service-Policy Provisioning
CORVUS	Cognitive Radio Approach for Virtual Unlicensed Spectrum
CPE	Consumer Premises Equipment
CPNP	Calling Party's Network Pays
CR	Cognitive Radio
CRAC/H	Computer Room Air Conditioning/Handling
CRC	Cyclic Redundancy Check
CRS	Cognitive Radio Services
CS	Circuit Switch
CSCF	Call Session Control Function
CSCI	Climate Savers Computing Initiative
CSI	Channel State Information
CT	Cooperative Transmission
CWC	Cooperative Wireless Communication
DAB	Digital Audio Broadcasting
DARPA	Defense Research Advanced Projects Agency
DCiE	Data Center infrastructure Efficiency
DCP	Data Center Productivity
DCS	Distributed Control Systems
DiffServ	Differentiated Service
DMA	Direct Memory Access
DMB	Digital Multimedia Broadcasting
DNS	Domain Name Server
DOCSIS	Data Over Cable Service Interface Specifications
DoS	Denial of Service
DoT	Department of Telecommunications
DRM	Digital Rights Management
DSA	Dynamic Spectrum Access
DSCP	Differentiated Service Code Points
DSL	Digital Subscriber Line
DSLAM	Digital Subscriber Line Access Multiplexer
DSL-F	DSL Forum
DSSS	Direct Sequence Spread Spectrum
DTH	Direct to Home
DTMF	Dual Tone Multi Frequency
DTT	Digital Terrestrial Television
EC	European Commission
EPC	Electronic Product Code

ERP	Enterprise Resource Planning
ESIF	Emergency Services Interconnection Forum
ESP	Encapsulated Security Payload
ETS	Emergency Telecom Service
ETSI	European Telecommunications Standardization Institute
EU	European Union
FCC	Federal Communications Commission
FDD	Frequency Division Duplexing
FEC	Forward Error Correction
FFD	Full-Function Device
FHSS	Frequency Hopping Spread Spectrum
FICCI	Federation of Indian Chamber of Commerce and Industry
FMC	Fixed–Mobile Convergence
FP	Framework Programme
FPGA	Field Programmable Gate Arrays
FSU	Flexible Spectrum Usage
FTTC	Fibre-to-the-Curb
FTTH	Fibre-to-the-Home
GDSN	Global Digital Synchronization Network
GERAN	GSM Edge Radio Access Network
GeSI	Global e-Sustainability Initiative
GHG	Green House Gases
GISFI	Global ICT Standardization Forum for India
GMDSS	Global Maritime Distress and Safety System
GPRS	General Packet Radio Service
GPRS	General Packet Radio Service
GPS	Global Positioning System
GSM	Global System for Mobile Communications
HAN	Home Area Networks
HDTV	High Definition Television
HFC	Hybrid Fiber Copper
HIP	Host Identity Protocol
HMI	Human Machine Interface
HSDPA	High-Speed Data Packet Service
HSS	Home Subscriber Server
HSUPA	High-Speed Uplink Packet Access
HTTP	Hyper Text Transfer Protocol
I-CSCF	Interrogating-Call Session Control Function
ICT	Information and Communications Technology

IEEE	Institute for Electrical and Electronics Engineer
IESG	Internet Engineering Steering Group
IETF	Internet Engineering Task Force
IKE	Internet Key Exchange
IM	Instant Messaging
IMS	IP Multimedia Subsystem
IMS-ALG	IMS Application Level Gateway
IMS-SF	IMS Switching Function
IMT	International Mobile Telecommunication
INC	Industry Numbering Committee
IoT	Internet of Things
IP	Internet Protocol
IP-CAN	Internet Protocol-Connectivity Access Network
IPSec	Internet Protocol Security
IPSO	IP for Smart Objects
IPTV	Internet Protocol Television
IPv4	Internet Protocol Version 4
IPv6	Internet Protocol Version 6
ISC	IMS Service Control
ISDN	Integrated Services Digital Network
ISIM	IMS Subscriber Identity Module
ISM	Industrial Scientific and Medical
ISO	International Organization for Standardization
ISP	Internet Service Provider
ISUP	ISDN User Part
ITS	Intelligent Transportation Systems
ITU	International Telecommunication Union
ITU-T	International Telecommunication Union-Telecom Standardization Sector
IVR	Interactive Voice Response
JNCC	Joint Network and Channel Coding
KPI	Key Performance Indicator
LAES	Lawfully Authorized Electronic Surveillance
LAN	Local Area Networks
LDPC	Low-Density Parity Check
LEA	Law Enforcement Agency
LI	Lawful Interception
LMDS	Local Multipoint Distribution Service
LMR	Land Mobile Radio

LoWPAN	Low Power Wireless Personal Area Network
LRIC	Long Run Incremental Cost
LTE	Long Term Evolution
M2M	Machine to Machine
MAN	Metropolitan Area Network
MBMS	Multimedia Broadcast Multicast Service
MB-OFDM	Multi-Band Orthogonal Frequency Division Multiplexing
MEF	Metro Ethernet Forum
MEMS	Micro Electro Mechanical Systems
MGCF	Media Gateway Control Function
MGW	Media Gateway
MIMO	Multiple Input Multiple Output
MMD	Multimedia Domain
MNP	Mobile Number Portability
MOU	Memorandum of Understanding
MPLS	Multi Protocol Label Switching
MRF	Media Resource Function
MRFC	Media Resource Function Controller
MTV	Mobile TV
NAT	Network Address Translation
NFC	Near Field Communications
NGN	Next Generation Networks
NIST	National Institute of Standards and Technology
NSF	National Science Foundation
NTIA	National Telecommunications and Information Administration
OCRA	OFDM Based Cognitive RAdio
OFDM	Orthogonal Frequency Division Multiplexing
OFDMA	Orthogonal Frequency Division Multiple Access
ONS	Object Naming Service
OSI	Open System Interconnection
PAN	Personal Area Networks
PDA	Personal Digital Assistant
PDU	Power Distribution Units
PLC	Programmable Logic Controller
PML	Physical Markup Language
PMRTS	Public Mobile Radio Trunk Service
ProWiN	Programmable Wireless Networking
PUE	Power Usage Effectiveness

QoS	Quality of Service
RAT	Radio Access Technology
RF	Radio Frequency
RFD	Reduced-Function Device
RFID	Radio Frequency IDentification
RISC	Reduced Instruction Set Computer
RRU	Remote Radio Units
RTOS	Real Time Operating System
SACFA	Standing Advisory Committee on Radio Frequency Allocation
SAML	Security Assertion Markup Language
SCADA	Supervisory Control and Data Acquisition
SDMA	Space Division Multiple Access
SDR	Software Defined Radio
SDRF	Software Defined Radio Forum
SEG	Security Gateway
SGIP	Smart Grid Interoperability Panel
SGW	Signaling Gateway
SICS	Swedish Institute of Computer Science
SIP	Session Initiation Protocol
SLF	Subscriber Location Function
SMTP	Simple Mail Transfer Protocol
SOA	Service-Oriented Architectures
SS	Spread Spectrum
TCO	Total Cost of Ownership
TCP	Transmission Control Protocol
TDD	Time Division Duplex
TDM	Time Division Multiplexing
TEC	Telecommunication Engineering Center
TGG	The Green Grid
THIG	Topology Hiding Inter-network Gateway
TIA	Telecommunications Industry Association
TISN	Trusted Information Sharing Network
TISPAN	Telecom and Internet Services and Protocols for Advanced Networks
TR	Technical Report
TRAI	Telecom Regulatory Authority of India
TSG	Technical Specification Group
UAS	Universal Access Service

UE	User Equipment
UICC	Universal Integrated Circuit Card
UMA	Unlicensed Mobile Access
UMTS	Universal Mobile Telecommunication Services
UMTS	Universal Mobile Telecommunication System
UNI	User Network Interface
URI	Uniform Resource Identifier
USD	Universal Service Directive
USN	Ubiquitous Sensor Networks
USO	Universal Service Obligation
USOF	Universal Service Obligation Fund
UTRAN	UMTS Terrestrial Radio Access Network
UWB	Ultra Wide Band
VLSI	Very Large Scale Integration
VoD	Video on Demand
VoIP	Voice over Internet Protocol
VoWi-Fi	Wireless Fidelity Enabled Voice over Internet Protocol
VSAT	Very Small Aperture Terminal
WAN	Wide Area Networks
WCDMA	Wideband Code Division Multiple Access
WEP	Wired Equivalent Privacy
Wi-Fi	Wireless Fidelity
WiMAX	Worldwide Interoperability for Microwave Access
WLAN	Wireless Local Access Networks
WPC	Wireless Planning & Coordination
WRC	World Radio Conference
WSN	Wireless Sensor Networks
XACML	eXtensible Access Control Markup Language
XML	eXtensible Markup Language

1

Introduction

Ramjee Prasad

Founder Chairman, GISFI and Director CTIF, Aalborg University, Denmark

Information and Communication Technology (ICT) is critical to improve the competitiveness of industry and to meet the demands of the society and economy.

India is playing an increasingly important role in the processes of globalization in the field of the ICT industry and the research and development thereof. This country is experience a massive uptake in ICT industrial developments. Huge state-funded initiatives have also been launched in many other Asian countries, in parallel with large-scale research and development activities in Europe and North America and along with advanced developments in the wireless IT sector and its associated markets. At the same time, international organizations such as the International Telecommunication Union (ITU-R) [1], the European Telecommunications Standardization Institute (ETSI) [2], the Association of Radio Industries and Businesses (ARIB) [3], and other fora (e.g., the Telecommunications Industry Association (TIA) [4]) are actively structuring the rapidly growing ICT sector through preparation work and standards.

The Global ICT Standardization Forum for India (GISFI) [5] is an Indian standardization body active in the area of ICT and related application fields, such as energy, telemedicine, wireless robotics, biotechnology. GISFI was established as an effort to create a new coherence and strengthen the role of India in the world standardization process by mapping the achievements in ICT in India to the global standardization trends.

With ICT services and applications developing fast in today's global economy and the ICT products and services becoming an integral part of our

R. Prasad (ed.), Future Trends and Challenges for ICT Standardization, 1–23.
© 2010 *River Publishers. All rights reserved.*

everyday life, it is of critical importance to create an environment which meets both the worldwide industry's needs and society's expectations; to promote the competitiveness of the industry while ensuring that all citizens can further benefit from the opportunities created by the advances in ICT [6].

Similar to other standardization bodies, GISFI is focused on balancing Indian and worldwide industrial requirements with expectations, and to provide public authorities with the standards they need to help them implement legislation in support of agreed policies.

This book comes in response to several major trends towards future ICT, such as the growing internationalization of ICT, the convergence of telecommunications and broadcasting, the need, importance and management of radio spectrum, the possibilities and challenges brought by the Internet of Things (IoT), and the development of future access technologies.

This chapter is organized as follows. Section 1.1 introduces the trends and challenges of ICT standardization and the importance of including the industrial and societal requirements and expectations of rapidly developing regions, such as India. Section 1.2 outlines major developments worldwide in the area of future access and spectrum technologies. Section 1.3 describes the role of convergence for the successful uptake of ICT and underlines the possibilities offered by advances in this area for providing true broadband services to end users, achieving energy efficiency of the teleinfrastructure and supporting the standardization effort. Section 1.4 is about the challenges and possibilities offered by the Internet of Things (IoT). In particular, this area has a main challenge of the interoperability between the large number of diverse devices and technologies. Section 1.5 describes the importance and challenges related to energy sustainable communication infrastructure. Section 1.6 gives a short summary of the Indian situation.

1.1 The Role of ICT Standardization

ICT is a major driver of competitiveness and currently represents one of the key industrial sectors [7]. Standardization is very important for providing a strong framework and conditions for ICT to contribute to economic growth and positive societal development.

Standardization is a voluntary cooperation among industry, consumers, public authorities and other interested parties for the development of technical specifications. Industry uses standards to support its competitiveness and to ensure acceptance of innovative solutions, which in turn increases interoperability and leads to market growth. The public authorities refer to standards in legislation, policies and procurement to achieve societal aims for safety, interoperability, accessibility, environmental performance, etc. While the industry can use any standards, the public authorities have a strong preference for, or even an obligation to use standards resulting from open, transparent and inclusive processes. Through referring to and promoting standards, the public authorities can help drive the competitiveness of industry and facilitate a competition for the benefit of the end users.

ICT standards development plays an important role for innovation, thus influencing the business investment decisions on research and development directions of this sector. As a source of the most up-to-date technical knowledge, the ICT standards broaden the knowledge base of the economy by integrating new technologies and research results harmoniously into the design and development process of new products and services. As a market instrument, the ICT standardization helps the business worldwide to create a common trading language. A prominent example is the Global System for Mobile Communications (GSM), which created a worldwide international market for mobile communications and was a driving force for the development of many mobile applications and technologies [8].

ICT standardization can ensure a steady fast-forward development of the ICT sector and overcome the challenges arising from the current trends and developments by developing an insight on dynamic business models that tie the technology, markets and consumers together by creating a platform for the participation of the various experts in critical discussions. ICT standardization creates a platform for the interaction with operators, regulators, manufacturers, Internet Service Providers (ISPs), infrastructure and software providers that can profit through the newly generated business.

1.2 Future Radio Access, and New Spectrum Technologies

The radio systems already being deployed are mainly evolutionary technologies (e.g. the cellular mobile systems such as GSM/General Packet Radio Service (GPRS) [9], Code Division Multiple Access (CDMA) 2000 [10], Universal Mobile Telecommunication System (UMTS) [11], time-division (TD)-CDMA, High-Speed Data packet Service (HSDPA)/High-Speed Uplink Packet Access (HSUPA) [12], and EV-DO). Research and development efforts, including standardization, are focused on both the evolutionary and revolutionary approaches (e.g., Third Generation Partnership Project (3GPP) long-Term Evolution (LTE) and LTE-Advanced [13]), the evolution of 3GPP2 EV-DO (Revision C or AIE), IEEE 802.16e and 802.16m Mobile

WiMAX [14], IEEE 802.20, and proprietary systems such as Flash-OFDM. Much effort has been put into standards (e.g., IEEE 802.15) that allow for efficient use of ICT technologies in our economic and social lives.

The realization of spectrally efficient radio interfaces can be achieved by use of technologies such as multiple input multiple output (MIMO) techniques, and in combination with adaptive modulation and coding, cognitive radio technologies, novel ways of channel and propagation modeling, interference modeling, and so forth. Enhancing the performance of radio interface technology can help to reduce spectrum demand, while accommodating more users. The spectral efficiency of MIMO transmission can be significantly increased if some level of channel state information is available at the transmitter, allowing the system to effectively adapt to the radio channel and take full advantage of the available spectrum. The main challenge is to make the channel state information (CSI) available at the transmitter CSIT). This can be achieved by conveying CSI as feedback information over the reverse link or requesting the CSI at the transmitter during receiving periods.

Adaptive modulation and coding techniques together with MIMO are the principal radio techniques to be considered when developing future radio systems. Low-density parity check (LDPC) and duo-binary turbo codes have been proposed and studied in adaptive coding and modulation schemes for next generation cellular systems. Such systems show only small performance differences when compared to schemes such as convolutional or block LDPC (BLDPC) codes, but can be competing forward error correction (FEC) schemes for larger block lengths. The choice of an optimal multiple-access scheme or a combination of schemes is very dependent on the performance requirements set up for the radio interface system. Adaptation to channel states, user requirements, and changing traffic patterns are only some of the considerations for the final choice. Other important considerations are that the radio interface system under design be able to interwork successfully with already existing or yet to be developed systems.

The best choice of frequency band for a mobile communications system depends on many different factors. Taking into account the physical nature of the fading radio channels, for mobile systems suitable spectrum should be as close as possible to the bands that are available for the existing mobile communication systems. For example, considering target peak data rates, target grade of mobility and target coverage range requirements of the future mobile communication systems, the maximum tolerable operating frequency is expected to be around 6 GHz, which is considered to be an upper limit for high mobility in the ITU-R [1]. However, bands greater than 6 GHz can be used for the new nomadic/local area wireless access.

Among the challenges related to spectrum, it is important to look at how the spectrum fits into the value chain and what are the benefits of promoting spectrum access, how to implement flexible spectrum, and another outlook on spectrum regulation. The goal of spectrum regulation and economic regulation should be the same: to pursue the long-term interests of the end users.

To support the demands of future wireless systems for high data rates and large user capacities, efficient use of the available spectrum resources is of great importance. Spectrum scarcity and bandwidth fragmentation lead the actual demand for transmission resources to often exceed the available bandwidth. Dynamic spectrum sharing and Flexible Spectrum Usage (FSU) are promising technologies to overcome this problem. The new spectrum allocated for mobile communications does not fully correspond to what is required for International Mobile Telecommunications (IMT)-Advanced systems. Therefore, the mobile operators in some countries might be forced to aggregate spectrum of two or more separated sub-bands for down- and uplink bands. Further, being able to cope with multiple bands will add some complexities both at the network managing and at the User Terminal (UT) level. The network needs to be aware of multiple bands and new scheduling and traffic engineering mechanisms need to be designed. The UT, on the other side, will need more advanced signal processing and multiple RF transceivers in order to transmit and receive on widely separated frequency bands.

In December 2004 3GPP started a feasibility study on LTE with an objective to develop a framework for the evolution towards a high-data-rate, low-latency and packet optimized radio-access technology [12]. The initial deployment of LTE is expected in 2009. Recently 3GPP has started aiming to further evolve LTE towards LTE-Advanced (LTE-A) to meet or exceed the IMT-A minimum requirements in terms of capacity, data rates and low cost deployments as well as its own requirements to further evolve the system.

Mobile communication is an important economic driver generating growth. Significantly improved transmission capabilities are increasingly required to support increased traffic originating from content-rich data services in order to connect people as well as machines to the information society. Frequency spectrum, as a scarce resource, requires efficient use and sharing techniques between different Radio Access Technologies (RATs) are important.

A set of frequency bands were identified by the WRC-07 for International Mobile Telecommunications (IMT) systems [1]:

- 450–470 MHz globally;
- 698–806 MHz in Region 2 and nine countries in Region 3;
- 790–862 MHz in Regions 1 and 3;
- 2.3–2.4 GHz globally;
- 3.4–3.6 GHz in a large number of countries in Regions 1 and 3.

The WRC'07 allocations are used as the starting point in the considerations of preferred spectrum use: the target is to define how to use the bands in the most efficient way. Within ITU-R standardization, for operation in paired/unpaired bands, a flexible spectrum usage between frequency division duplex (FDD) and time division duplex (TDD) operation should be defined, as well as static and dynamic guard bands.

Flexible sharing of licensed and unlicensed bands can lead to increased spectral efficiency by allowing the use of unused spectrum of other systems or cellular operators. This is especially important for ubiquitous future high data rate services in next generation systems, whose requirements cannot be met with traditional approaches, which suffer from their lack of flexibility. A very important requirement for flexible spectrum use is that the interference to the other systems must remain low. Being able to cope with multiple bands will add some complexities both at the network managing and at the UT level. The network needs to be aware of multiple bands and new scheduling and traffic engineering mechanisms need to be designed.

Spectrum aggregation can appear when the operator's dedicated band is not continuous but is split in two or more parts. In addition, spectrum aggregation can happen in scenarios in which an operator accesses both a dedicated band and a spectrum-sharing band, which is separated in frequencies from the dedicated operator's band. Being able to cope with multiple bands would increase the flexibility and the efficiency of spectrum utilization, but will also add some complexities both at the network managing and at the UT level.

Lately, a lot of discussion and work has been devoted to the topic of liberalization of the spectrum and its technology-neutral allocation. Technology-neutral spectrum allocation and liberalization principles enable the spectrum holders to change the use of that spectrum, e.g. by migrating spectrum used for 2G and 3G systems for the use by IMT-Advanced systems, to lease or, even to sell it on secondary spectrum markets.

A related issue is how to populate the spectrum made vacant by the switch of analogue terrestrial TV broadcasting in the frequency band 470–862 MHz to digital TV. Since the processing of digital data enables a more efficient use in terms of required bandwidth, a considerable amount of frequency spectrum can be released from broadcasting use leading to the digital dividend approach, which allows the launching of IMT commercial systems in this band.

Femto-cells, i.e. miniature base-stations that are deployed in an operator-owned spectrum and are based on the same cellular standards as macro-cells, pose several important regulatory challenges. For example, the conditions under which users are allowed to carry the Femto cell outside of their home area. Also, an important issue is access for emergency calls to the closed access Femto cells of mobile users that are not subscribed to that Femto cell.

The placement of Femto cells has a critical effect on the performance of a previously deployed macro network. While operating Femto cells on a dedicated frequency band is a simple solution, co-channel operation with existing macro-cell is technically challenging, although it is a rewarding approach for operators. One of the main problems yet to be solved is that Femto cells and macro cells could suffer severe co-channel interference from each other if cell planning or spectrum management is not appropriately considered. In practice it might not be possible to perform cell planning because e.g. due to the fact that the Femto base stations (BSs) are likely to be deployed by the subscribers, it seems difficult for the operator to be able to pre-configure the Femto BSs such that one minimizes the interference to other Femto BSs to suit the deployment location. It will be difficult to estimate beforehand the level of interference that a Femto BS could practically cause to the other Femto BSs. Hence careful consideration is need to determine how the Femto BSs will coexist with each other [15]. So, many parameters need to be automatically configured considering local conditions such as distance from co-channel macro cells or received signal power from other neighboring Femto cells when Femto BSs are plugged in [16]. Power control may ensure low impact on the macro cells, but it involves additional complexity for Femto base stations [17].

A main challenge is to combat the interference, which can have several scenarios. Figure 1.1 shows an example of Femto-to-Femto interference. A technically demanding possibility related to intelligent spectrum management is "network-sharing". In this modality, the whole radio access system (from the fiber connection towards the antenna) is shared among several operators. A scenario for this sharing is: large, lightly populated areas, which individually do not provide enough revenue to cover the deployment cost. Operators may share a BS, and feed the BS with own fiber connections to their respect-

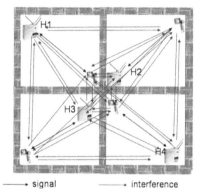

——→ signal   ——→ interference

Figure 1.1 Example of Femto-to-Femto interference

ive core networks. In other situations, depending on the agreement, operators may split the coverage areas, being responsible for the SLA and QoS in their respective area and providing virtual coverage to the other one by means of roaming services transparent to end user. Operators already have constructed test-beds and set up operative pilots for some of these concepts. This concept is very much related to developments in the area of convergence, which would be discussed in more detail in Section 1.3.

1.3 Convergence

Telecommunication networks architectures are changing to meet new requirements for a number of services/applications (Broadband, IP, Multimedia, mobile, etc.). New generation equipment (soft switches, databases, service controllers, new protocols and interfaces, etc.) and new call/mix traffic cases are being introduced in the networks.

The selection of new technology depends on the given infrastructure availability and depends on the specifics of the corresponding rural and urban areas. This in turn affects the consequent network planning process. In particular, in developing countries, the needs may be substantially different in urban and rural areas.

Even when ICT-based development is planned, such differences may persist in urban and rural areas, and within urban areas. Accordingly, the infrastructure and technology requirements will differ. Thus, no single technology can meet all traffic, market and operational requirements. There will be no clear-cut optimum technology but rather a number of technologies

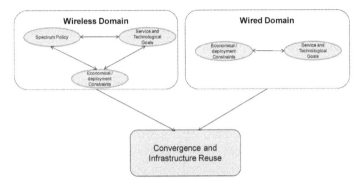

Figure 1.2 Convergence of wired and wireless towards a common digital infrastructure (adapted from [19])

with different characteristics. In choosing technologies for a new or existing network, a very wide range of factors needs to be considered.

A high quality digital infrastructure enables the extensive use of home workspaces and videoconferencing thereby decreasing the commuting needs – local and global. Fiber to the home (FTTH) in combination with a reliable wireless network infrastructure is one enabler of high-quality ICT services.

It can be said that convergence was spurred by interdependent technical, regulatory and economical/deployment trends related to the vision of what will be the infrastructure for next generation networks. This is shown in Figure 1.2.

Convergence is about the provision and deployment of true broadband wireless access, constraining the technical solutions by capital and operational expenditures and ensuring the required flexibility to allow upgradeability and reconfigurability in order to match the dynamics created by the different players in the field and the needs of the end users for more and more bandwidth, and sophisticated services.

The provision of broadband services to everyone is considered one of the key components for enabling the so-called information society. Traditionally, the delivery of broadband connections to the end user has been targeted through the deployment of optical fibre leading to the concepts of Fibre-to-the-Curb (FTTC) and Fibre-to-the-Home (FTTH). However, during the last decade, the impact of wireless telephony and Wireless Local Access Networks (WLANs) has been such that the liberation of a physical connection created in the end user a new sense of freedom and autonomy in his relation with communications [18].

Nevertheless, the provision of the high bit rates ($\sim$1 Gbit/s) that could be envisioned of interest for the end user and more easily provided with fixed optical connections still represents an enormous challenge to the wireless community. As a consequence of the technical trends, such as to provide higher bit rates, Quality of Service (QoS), reconfigurability, and adaptability, the complexity of wireless systems is rapidly increasing. Therefore, if one stays with the current paradigm of a cellular architecture, the base stations will be more and more complex, and this coupled with the cell size reduction required to enhance the overall system capacity, will imply that the deployment of these sophisticated base stations will be a major factor in the overall network cost. Cell site location is critical. The location must have enough space to support the antenna and any ancillary equipment should have available power, easy access for technicians, and be free of obstructions that would block coverage. The number of sites meeting all, or at least most of, these criteria gets smaller everyday. Often, even if technically suitable sites are available, the site-owner's terms of use are unfavorable and push providers toward less desirable sites. Of course, the less desirable sites create new coverage issues that force the provider to accept spotty coverage or find and populate an additional site. In addition to the decreasing number of acceptable sites, service providers face opposition from the same communities demanding better coverage. Many of these communities place severe limitations on the construction of wireless sites. As a result, stringent and ultimately costly requirements for location and appearance restrict available locations. With conventional sites limited and expensive, a new approach is required to fulfill the promise of anytime, anywhere access. One solution is to simplify the Base Stations (BS) as much as possible, reducing them to simple RAUs that just transmit/receive radio signals and are connected through optical fiber to a central unit where all the processing is performed, bringing significant advantages both in terms of reduced installation and operational (maintenance) costs.

The infrastructure owners operators own the converging communication infrastructure (e.g., fiber, WLAN, WiMAX, UMTS, broadcast systems, IMT-Advanced systems, etc.). The technical network operators rent the capacity in this infrastructure that is then combined in packages so that bit pipes fulfilling certain QoS requirements can be offered to virtual network operators who can sell the service to the subscribers. This is also the vision of the wireless ecosystem as shown in Figure 1.4.

In a converging scenario, the wired and wireless technologies should be integrated as cooperating and not competing means of access and the design of the converged network can be done in the same manner [20]. Doing so, achieves the full synergy effects in the deployment with shared locations, trenches, tubes or even fibers. Besides, for various reasons, a general and full quantification of the deployment of radio access systems is impossible to perform. Certain deployment components are operator-specific as they depend, e.g., on the contractual relationship between the hardware manufacturer and the operator. Others depend on the regulatory and legislative environment in the country of deployment.

A combined and optimized deployment network planning is a must for enabling a converged infrastructure complying with all requirements of the modern society. The network planning process must be optimized to include the aspect of energy-efficient communication networks. A ubiquitous heterogeneous network environment will integrate wireless and converged wired/fixed network segments and by default implies the provision of services over multiple domains.

Thus, convergence must be considered hand-in-hand with technologies such as IoT and green ICT, which is an important part of the road towards successful deployment of novel ICT technologies.

1.4 Internet of Things

The general meaning of the term Internet of Things (IoT), refers especially to everyday objects, that are readable, recognizable, locatable, addressable, and/or controllable via the Internet – regardless of the type. Everyday objects include not only the common electronic devices, and not only the products of higher technological development such as vehicles and equipment, but things such as food and clothing. An IoT scenario is characterized by heterogeneity found in the types of participating nodes, networks and type of data exchange. This creates a complicated scenario in relation to decision making, management and reliability.

The autonomic functionality of nodes is a feature also known as self-x (e.g. self planning, self configuration, self optimization, self tuning, self healing, self maintenance and self management). For example, the need to decrease network maintenance costs and increase the network reliability calls for automatic fault recovery and reconfiguration of components which could be achieved using self-healing and self-configuration. Another example is the need to increase the network or device performance through self-optimization actions. This case may arise when base stations decide to perform self-optimization actions based on their own policies (e.g. minimize energy consumption, or maximize the supported traffic in an area).

The need for self-*x* characteristics, instead of using centralized solutions, mainly stems from the explosion in the number of mobile devices and nodes as well as the complexity that heterogeneous environments tend to create. Self-management functionality enhanced with cognitive mechanisms is expected to address these problems efficiently. Cognitive mechanisms monitor the current network conditions and are able to plan, decide and act on those conditions. Moreover, they can learn from the consequences of their actions. This knowledge is used to improve the effect of similar decisions in the future operation of the network. Cognitive networks cover all layers of the protocol stack and actually form a "knowledge plane" that can span over layers vertically while making use of cross-layer designs.

The first step will be for a node to monitor its "environment" (i.e. collect useful information from the signal strength, current battery levels, etc.) and detect any notable changes. Autonomous decisions would be made based on context information, policies and any previous knowledge. Thus, a node is able to select the most appropriate radio access technology for its communication. This decision may require a protocol reconfiguration and even the self-organization of several nodes into clusters that may serve better their needs (e.g. in an attempt to bypass a network failure in an area). The result of the aforementioned actions is then evaluated and the produced knowledge is stored for future use. This type of functionality can be a real advantage especially in critical scenarios where past knowledge can minimize the reaction time of the nodes and satisfy their communication needs in best available way.

Further, such an autonomous network control and management system, which can learn from observing how content is accessed by region and from content popularity, requires new middleware in order to determine how best to distribute digital content and how to best route a request to the closest server.

Considering also the mobility of the user and the support of various types of services (e.g., streaming), the user may wish to switch endpoints at certain points in time, e.g. video or voice streams may be switched from handheld and mobile terminals, to the user's home entertainment system as soon the user reaches his home. Or the system may wish to switch the streaming server, as the currently used one is being overloaded by many users or latency requirements cannot be met at the current server, but with other servers it may be possible. Such mobility scenarios require support from the middleware in terms of control messages for flow synchronization and session management, but also in terms of detection of when, where and how to shift the service interaction. Different degrees of vertical integration or separation can allow a number of sharing opportunities.

A major future technological trend will be to embed communication and computation processes within miniature objects. This will lead to the creation of situated and cooperating "smart" objects with sensing and actuating capabilities able to perceive and control their environment. Such concealed smart devices are the basic building blocks for environments with embedded intelligence and will link the digital world of the Internet with the real world.

Because of the vast number of distinct, wireless networked sensing and actuating devices and their rather limited resource capacity, isolated entities can no longer perform this interaction efficiently or reach the required distributed control objectives. As a result, cooperation between individual entities is a necessity, for energy-efficient, distributed monitoring and control tasks in large-scale systems. Cooperating objects are fundamental to embedded systems where everyday objects can be endowed with computational capacity; wireless communication and advanced control techniques.

The central aim of systems of cooperating objects will be to develop ambient intelligence technology that can "glue" together diverse objects to enable seamless environments for computing, communication and service delivery. Such technology should be open, distributed and scalable, naturally integrating heterogeneous devices ranging from tiny actuators to large computers. It will combine architectures, operating systems, middleware, programming models and tools to support location and context sensitivity, timely reactivity and pro-activeness, real-time adaptation and security. Objects should be seen as encapsulating entities, mixing hardware and software in any proportion and at any level of complexity. So, an object may be as simple as a sensor, as complex as a portable device, or even an entire car or building.

Figure 1.3 shows a roadmap envisioned for the development of IoT and the main drivers it requires. It highlights the timing, features, and applications of significant technology milestones that would be necessary for developers of this technology to achieve if successful (equivalent to commercial) application – and possible disruption – is to occur by 2025.

There are two distinct modes of communication in the Internet of Things that can be defined as follows:

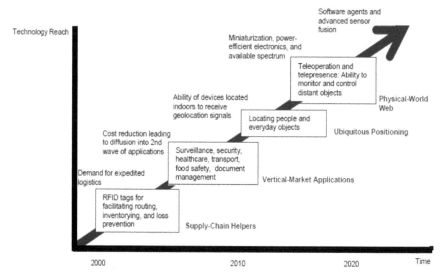

Figure 1.3 IoT roadmap (Source: SRI Consulting Business Intelligence)

- Thing-to-person (and person-to-thing) communications encompasses a number of technologies and applications wherein people interact with things and vice versa, including remote access to objects by humans, and objects (sometimes called "blogjects") that continuously report their status, whereabouts, and sensor data.
- Thing-to-thing communications encompasses technologies and applications wherein everyday objects and infrastructure interact with no human originator, recipient, or intermediary. Objects can monitor other objects, take corrective actions, and notify or prompt humans as required. Machine-to-machine communication is a subset of thing -to-thing communication; but machine-to-machine communication often exists within large-scale IT systems and so encompasses things that may not qualify as "everyday objects".

IoT can be seen as a strategic path for increasing the dynamics of the telecom market by creating access to real-time information anytime, anywhere. It is enabled by the advances in technology such as RFID, sensors and actuators, embedded systems, and so forth. The best way to characterize IoT from other technologies is that it is an environmentally aware Internet technology where objects do not necessarily have Internet addresses, and where objects can be connected at a different range by means of various technologies, and have different capabilities.

Other important enablers for the future of IoT are technologies, such as machine-to-machine interfaces and protocols of electronic communication, embedded micro-controllers, wireless communication, energy harvesting technology, positioning technology, software platforms.

1.5 Green ICT

Green IT is one of the main topics dealt with by governments and product makers. Novel ICT solutions are being proposed for support of the electricity networks (e.g., smart electricity distribution networks), as the basis for the development of tools leading to energy-efficient solutions in various spheres of the industrial and every-day life (e.g., monitoring and control systems able to optimize the local energy generation-consumption). The challenge of designing ICT networks and communications to be more energy and cost efficient has many aspects. For example, base stations powered by wind turbines are an approach to energy efficiency and sustainability for the wireless communication infrastructure [21]. Advanced technologies relying on distributed architectures are a means to reduce the information exchange and release resources without the need for new deployments. Here, it should be mentioned that although relays can give the benefits of increased coverage and other benefits of a distributed architecture [22] a very dense deployment might raise environmental concerns. Network planning, including the reuse of existing infrastructure, therefore, is an essential aspect of developing energy-efficient and sustainable (e.g., "green") communications.

The pattern of commuting in small cities and rural areas can be changed to benefit the overall reduction in CO_2 emissions by designing a digital infrastructure able to support a high quality of communications and services for the home and the office. A high quality digital infrastructure enables the extensive use of home workspaces and videoconferencing, thereby decreasing the commuting needs – local and global. Fiber to the home (FTTH) in combination with a reliable wireless network infrastructure is one enabler of high-quality ICT services [19].

The introduction of advanced radio technologies (e.g., flexible spectrum sharing) in a converging communication environment calls for investment decisions in a changing and complex wireless ecosystem.

An example of a wireless eco-system is shown in Figure 1.4.

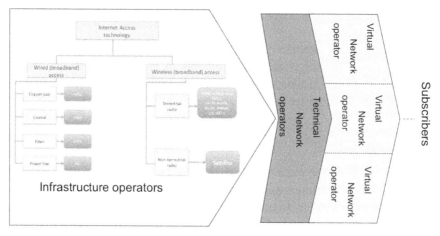

Figure 1.4 An example of a wireless eco-system (adapted from [23])

Technological innovations have brought many transformations to society, including the energy sector. Every day, ICT plays a part in the production and use of electricity, from the operation of power plants to electricity transport. A recent trend is the research and development effort of using ICT to model, simulate and optimize these processes.

The manufacturing industry is facing a number of challenges that can be summarized as follows:

- mass customization,
- last-minute changes,
- cross-organizations processes,
- a very heterogeneous world.

The above challenges shape the vision for future manufacturing processes and these arise from the following trends: agile manufacturing, smart objects penetration, and service-oriented architectures.

Thus, there is the need for an increased flexibility and agility in the manufacturing process, which can be delivered, among others, by dynamic reconfiguration and composition of services and by integrating lower layer events to higher layer enterprise applications.

1.5.1 Monitoring and Control

Today's progress in manufacturing is determined by the advances in control engineering and the multi-disciplinary cooperation and synergies with computing, communications, sensors, cognition etc. This area becomes more and more interlinked with advances in software development, convergence (see Section 1.3) and IoT (see Section 1.4). The current research and development activities related to the area of green ICT address engineering technologies for large-scale, distributed and cooperating systems for monitoring and control, the development of architectures, hardware/software integration platforms and engineering methods for distributed systems composed of heterogeneous networked smart objects. Research challenges include, for example, methods and algorithms to support spontaneous ad-hoc cooperation between objects; network-centric computing with dynamic resource discovery semantics that allow object/service definition/instantiation; lightweight operating systems and kernels [24].

On the other hand, modern datacenters consume huge amounts of electricity and usage is increasing at a rapid pace. Energy consumption has accelerated as applications move from desktop computers to the Internet and as information gets transferred from ordinary computers to distributed "cloud" computing services.

Businesses can be boosted by creating technologies and platforms allowing for the extraction and the evaluation of information from the environment related to energy consumption (e.g., per building). Further, the supporting technology for collecting and delivering this information, as well as the decision mechanisms can be made in itself energy-aware. The problems that can be currently identified for businesses relate to: lack of visibility for energy consumption and the need for transparency. An important issue to consider is that the information about CO_2 emissions can be a very useful indicator.

Open research issues are related to the ways the data from infrastructure can be fed to services: for example, it could be useful to combine a set of services to the task one wants to achieve.

1.5.2 Platforms for Energy-Efficient Applications

One direction of research is towards Web-based service-oriented architectures (SOAs), which support the connectivity of smart devices (i.e., machines on shop-floor) to high-level backend systems (e.g., ERP services) [25].

Such architectures can be applied to smart metering and control of the house and of systems from outside.

Smart metering can be applied for everything and it does not have to be static, it can also be located outside of the home. A mobile device can also be a meter.

The need is for network-based services, whereby horizontal collaboration is a benefit for the creation of new applications.

Key performance indicator (KPI)-driven optimization of dynamic processes that can be changed over time could lead to cost effectiveness. Such an approach allows for each process to have different tasks and each task can be distributed to different execution processes.

Social networks could also be involved in measuring and informing about the energy consumption for comparisons and sharing of energy use, enabling a dynamic discovery of energies. In general, monitoring is very beneficial for increasing visibility. Real-time simulations can be used for obtaining the control messages.

Monitoring and control in relation to support of green ICT is an important scientific and technological field and market sector. It is based on rigorous sensing/monitoring, modeling, decision making, control and actuation and optimization. Sensor/actuator networks and wireless sensor network (WSNs) constitute a critical enabling technology, and cooperation amongst smart objects is an important extension thereof.

Research in the area of monitoring and control can be addressed by focusing on large-scale deployments and complex systems. Large-scale/complex deployments constitute the next major challenge, extending from instantiation to management and maintenance to updating. These should enable optimal operation of large-scale dynamic systems through proactive process automation systems applicable across several sectors, going far beyond what current supervisory control and data acquisition (SCADA) and distributed control systemsdistributed control systems (DCS)/programmable logic controller programmable logic controller (PLC) can deliver today. Pro-activeness requires novel predictive models. The architectures should facilitate re-use, enable Quality of Service (QoS), and reduce the reconfiguration effort. Standardization of monitoring and control systems in industrial environments is recognized as a key factor towards a green ICT society [24].

In general, the ongoing research and development in the area should strengthen the competitiveness of the industry supplying monitoring and control systems through next generation process automation products that are superior in terms of functionality, accuracy, dynamic range, autonomy, reliability and resilience. Further, it should lead to improved ease-of-use and simplified operation and maintenance of monitoring and control systems, also for non-experts.

Application-specific adaptable communication technologies enabling the real deployment of smart WSN in large-scale, performance-critical application fields like automotion and aeronautics is another area enabling green ICT. Application scenarios in these fields entail a complex and heterogeneous set-up, in which different nodes require or provide different performance. An important aspect is the support and the compatibility with other (e.g., vehicle) networks, both new and legacy. Energy harvesting techniques are key to enabling these scenarios, the identification of an application scenario that stresses the network; the enhancement of an existing functionality, the degree of reliability of sensors in an automotive environment, and wireless data fusion are only some of the aspects of the work.

The development of new hardware and software platforms enabling distributed optimal execution and scalable performances is another enabling area. In this particular case, middleware architecture is needed for the support of the system auto-configuration through dynamic resources discovery and management [26].

1.5.3 Embedded Systems and Cooperating Objects

Reducing the complexity and facilitating the access and control of embedded sensors is another enabler towards energy efficiency. This can be very beneficial to support of middleware platforms that could lead to thousands of new applications in a range of industries. Such a platform can access the functionality, but hide the underlying complexity of embedded sensors, making development and deployment of new services a snap. An efficient way to treat information from current embedded systems is by first building the position of the obstacle and conveying the information to the infrastructure, then building the applications on top of other applications [27]. A specific platform can be built to take the difficultly out of developing new applications for the existing embedded sensors. It would act as an interface between the designers and the electronics. The scope of research in this direction are the wireless cooperating objects and the interactions thereof, new routing protocols to optimize mobility and a framework for use of this information with already existing information.

A way to create smart services is making the information in the digital world available to the physical world. The common target is to enable and maintain cross-industry interoperability, to foster innovation while maintaining value of existing legacy and to create new user interaction and interface concepts to enable users to benefit from smart environments. Through specific software development multiple applications from multiple vendors can be allowed to share ad-hoc information across numerous domains, enabling cross domain and cross platform interoperability [28].

Producers of devices and components are increasingly facing the need for networking their own and complementary products in order to provide higher value-added solutions for their customers or because end-user-centered demands require much more focus on intelligent solutions where the complexity is hidden behind user-friendly interfaces in order to promote inclusion. Given the enormous amount of heterogeneous devices, sensors, and actuators with embedded systems already existing in the market, the diversity of the producers and manufactures, the different clock speed of technology deployment (from several decades to some months), there is a very large need for technologies and tools that easily can add, implement and exploit the intelligence embedded in the devices. The goal for the developers is to be able to build cost-efficient Ambient Intelligence (AmI) systems with high performance, high reliability, reduced time to market, and faster deployment and still build on the assets of the installed base. Access to device features is radically different among the various devices, ranging from human-only access (e.g. screens and buttons), to exclusive machine-to-machine communication over standard protocols and possible wireless transports. An example of AmI is shown in Figure 1.5.

This wide array of access mechanisms is acceptable for most developers of the device-based intelligent solutions, such as skilled system integrators, who need to access only a few devices. It is much more difficult to build solutions based on a large number of devices from different manufacturers with heterogeneous access protocols that for the most remain proprietary or unknown. The current status on the market thus makes it almost impossible for existing devices to communicate and exchange information, without the direct involvement of the device manufacturers in one way or the other. The first challenge is thus to allow for the seamless access to the features of many devices, regardless of its manufacturer, technology, interfaces, location, communication mechanism, etc. and to provide intelligent and secure interoperability. Applications should adapt to changing local and global sets of accessible sensors and actuators, and must piece together partial states of internal and location-determined information. When an end-user moves around interacting with any device in either the private or public space, the right information must follow their migration from different locations in changing surroundings.

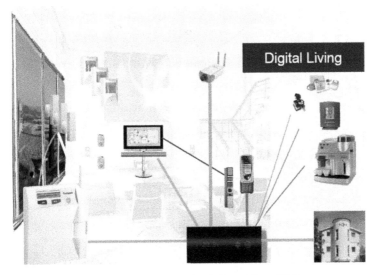

Figure 1.5 AML scenarios (adapted from [27])

Another challenge is thus to develop a framework for secure, trust-worthy communication among networked embedded systems and supporting self-adaptive interplay of different components, not only sensors but also controlling components and actuators.

1.6 Brief Indian Scenario

India is currently powered by rapid developments in IT, telecommunications, the World Wide Web, world-class entrepreneurs and a deep reservoir of technology manpower equipped with state of the art tools and workflow. At the same time, the communication technologies in India and worldwide are experiencing a rapid uptake in all spheres of society. For example, from twenty odd million fixed-wire connections in the early nineties, India today boasts close to four hundred million telephone subscribers (fixed wire and mobile combined) and the market is growing at a fast pace; in the past two years, an increase of mobile subscribers by almost ten million every month has been reported [29].

References

[1] International Telecommunications Union, ITU, at `www.itu.int`.

[2] European Telecommunications Standards Institute (ETSI) at `www.etsi.org`.

[3] Association of Radio Industries and Businesses (ARIB), at `www.arib.or.jp`.

[4] Telecommunications Industry Association at `www.tiaonline.org`.

[5] Global Standardization Forum for India (GISFI) at `www.gisfi.org`.

[6] G. Verheugen, The Specific Policy Needs for ICT Standardisation, Brussels, Belgium, February 2008, at `http://ec.europa.eu/enterprise/newsroom`.

[7] European Commission, White Paper, Modernising ICT Standardization in the EU-The Way Forward, July 2009, Brussels, Belgium, `http://ec.europa.eu/enterprise/sectors/ict/standards/`.

[8] European Commission, Communication to the European Parliament and Council: On the Role of European Standardisation, October 2004, Brussels, Belgium.

[9] GPRS technology, `www.gsmworld.com/technology`.

[10] R. Prasad, *CDMA for Wireless Personal Communicaitons*, Artech House, 1996.

[11] UMTS, `www.umts-forum.org`.

[12] 3GPP Technical Report, Further advancements for E-UTRA Physical layer aspects, February 2009, at `www.3gpp.org`.

[13] Long term Evolution (LTE) and LTE-Advanced, `www.3gpp.org`.

[14] Standard IEEE 802.16e, `www.ieee802.org/16/tge`.

[15] S. Kumar and N. Marchetti, IMT-Advanced: Technological Requirements and Solution Components, in *Proceedings of Wireless VITAE*, May 2009, Aalborg, Denmark.

[16] L.T.W. Ho and H. Claussen, Effects of user-deployed, co-channel femtocells on the call drop probability in a residential scenario, in *Proceedings of IEEE International Symposium on Personal, Indoor and Mobile Radio Communications*, pp. 1–5, September 2007.

[17] T.-H. Kim and T.-J. Lee, Throughput enhancement of macro and femto networks by frequency reuse and pilot sensing, in *Proceedings of IEEE International Performance, Computing and Communications Conference (IPCCC)*, pp. 390–394, December 2008.

[18] FP7 EU-funded project FUTON, at `ftp://ftp.cordis.europa.eu/.../fp7/.../projects-futon-project-factsheet-fv_en.pdf`.

[19] ICT FP7 EU-funded Project FUTON, at `www.ict-futon.eu/`.

[20] M.T. Riaz, R.H. Nielsen, J.M. Pedersen, N.R. Prasad, and O.B. Madsen, A framework for planning a unified wired and wireless ICT infrastructure, *Wireless Personal Communications*, DOI 10.1007/s11277-009-9720-5, 2009.

[21] J. Gozalves, White spaces, *IEEE Vehicular Technology Magazine*, 4(2):4–11, June 2009.

[22] IST FP6 EU-funded Project WINNER, at `www.ist-winner.org`.

[23] R.H. Nielsen, A. Mihovska, and O.B. Madsen, Energy-efficient deployment through optimisations in the planning of ICT networks, in *Proceedings of WPMC'09*, Sendai, Japan, September 2009.

[24] J.M. Pereira, The contribution of monitoring & control to energy efficiency: From a component to a system perspective, October 2008, `http://cordis.europa.eu/fp7/ict/necs/events-20081021_en.html`.

[25] S. Karnouskos, Monitoring and control for future energy-aware business processes, in *Proceedings of Third Concertation Meeting of Components, Systems and Engineering*, FP7 EU-funded Projects, Brussels, Belgium, 7 October 2009.

[26] T. Herndl, Cooperative hybrid objects sensor networks, http://cordis.europa.eu/fp7/ict/necs/events-20091007_en.html, October 2009.

[27] M. Serrano, Wireless embedded middleware in mobility applications, in *Proceedings of the Second EMMA Workshop*, Newcastle, UK, March 2008.

[28] P. Soininen, The smart-M3 concept, in *Proceedings of NOTA Conference*, San Jose, California, USA, October 2009.

[29] ICT Landscape-India, at http://www.euroindia-ict.org/index.php/ict-landscape-india.

2

Radio Frequency Spectrum and Regulation

K. Sridhara[1], P.K. Garg[2], Ashok Chandra[3] and P.S.M. Tripathi[4]

[1]*Ex Member, Technology ICT (Government of India) and Director General, GISFI*
[2]*Former Wireless Advisor, Government of India*
[3]*Wireless Advisor, Government of India*
[4]*Wireless Planning and Coordination (WPC), Government of India*

This chapter elaborates on the importance of radio frequency spectrum as a scarce natural resource which is under immense pressure due to the forthcoming deluge of data to be transferred on the network.

The first section of this chapter discusses various aspects of the spectrum itself and the need of managing it, and access technologies. The need for more dynamic access technologies in context of the evolution of Cognitive Radio Technology is discussed in Section 2.2. Section 2.3 discusses the Indian scenario with regard to spectrum issues and envisions regulatory aspects and policy making. This section also elaborates on the cognitive radio and spectrum management in that context. Key challenges and future research direction is envisioned in the fourth section. Section 2.5 discusses the framework evolved and some broad thinking about the cognitive radio as the future of wireless communication technology. Section 2.6 elaborates on spectrum issues for LTE and WiMAX technologies since these are on verge of deployment all over the world.

2.1 Radio Frequency Spectrum and Management

It is envisioned that 7 billion people will be served by 7 trillion electronic devices which will be networked and interacting. In this scenario the fre-

R. Prasad (ed.), Future Trends and Challenges for ICT Standardization, 25–74.
© 2010 *River Publishers. All rights reserved.*

quency spectrum which is a fixed asset poses a challenge for engineers, scientists and regulators all over the world. Spectrum management strategy has to be devised on the technological as well as policy levels to meet this challenge.

2.1.1 Need for Spectrum Management

The radio frequencies are useful for transmission of signals over long distances, and their propagation is in accordance with the laws of physics. The group of radio frequencies is called the RF spectrum (Table 2.1). It is a limited, natural resource that is available equally to all countries. The RF spectrum starts from 9 KHz and the upper limit extends up to 3000 GHz, above which lie the optical frequencies [1]. Theoretically the spectrum appears to be vast. However, the limitation of RF spectrum results from the availability of technologies for utilizing various frequency bands for different services and applications for which various telecom systems are to be deployed (Table 2.2). This limitation further increases when the availability of cost effective equipment for particular service(s) and applications is considered. At present, the spectrum up to approximately 40 GHz only is used for commercial applications [2] and up to 100 GHz for scientific/strategic usages. (Equipment costs are high for commercial usage.) Technologies for utilizing higher frequency bands are being developed. The spectrum scarcity also comes from policies that grant licenses which in turn offer exclusive access rights for some users to parts of the spectrum. These policies are again motivated by technical issues like interference and electromagnetic compatibility as well as by domestic welfare and national security concerns.

In the International Telecommunication Union (ITU) parlance, broadcasting, aero-mobile & maritime-mobile, communications for defense, police and other Para-military and security agencies, radio astronomy and radio amateurs, etc., are also part of "Telecommunications", besides public telecom services. There are more than 40 different radio communication services, which make use of different portions of the RF spectrum [3,4]. Examples of these radio communication services are fixed radio communication services, Mobile radio communication services, sound and television broadcasting services, satellite communication services, maritime and aeronautical mobile services, radio-determination (RADAR) service, etc. Each radio communication service has numerous applications. For example, public mobile (cellular) telephone systems, mobile wireless systems of police, fire fighting agen-

Table 2.1 Frequency bands and bandwidths used by different technologies worldwide

Uplink (UL)	Downlink (DL)	Duplex	Channel Bandwidths	Alias	Region
1920–1980 MHz	2110–2170 MHz	FDD	5, 10, 15, 20	UMTS IMT, "2100"	Japan, Europe, Asia
1850–1910 MHz	1930–1990 MHz	FDD	1.4, 3, 5, 10, 15, 20	PCS, "1900"	United States, Latin America
1710–1785 MHz	1805–1880 MHz	FDD	1.4, 3, 5, 10, 15, 20	DCS 1800, "1800"	Finland, Hong Kong
1710–1755 MHz	2110–2155 MHz	FDD	1.4, 3, 5, 10, 15, 20	AWS, "1.7/2.1 GHz"	US, Latin America
824–849 MHz	869–894MHz	FDD	1.4, 3, 5, 10	Cellular 850, UMTS850	US, Australia
830–840 MHz	875-885 MHz	FDD	5, 10	UMTS800	Japan
2500–2570 MHz	2620–2690 MHz	FDD	5, 10, 15, 20	IMT-E, "2.5 GHz"	EU
880–915 MHz	925–960 MHz	FDD	1.4, 3, 5, 10	GSM, UMTS900, EGSM900	EU, Latin America
1749.9–1784.9 MHz	1844.9–1879.9 MHz	FDD	5, 10, 15, 20	UMTS1700	US, Japan
1710–1770 MHz	2110–2170 MHz	FDD	5, 10, 15, 20	UMTS,IMT 2000	Brazil, Uruguay, Ecuador, Peru
1427.9–1452.9 MHz	1475.9–1500.9 MHz	FDD	5, 10, 15, 20	PDC	Japan (Softbank, KDDI, DoCoMo)
698–716 MHz	728–746 MHz	FDD	1.4, 3, 5, 10		
777–787 MHz	746–756 MHz	FDD	1.4, 3, 5, 10	Verizon's 700 MHz Block C	
788–798 MHz	758–768 MHz	FDD	1.4, 3, 5, 10		
704–716 MHz	734–746 MHz	FDD	1.4, 3, 5, 10	AT&T's 700 MHz Block B	
1900–1920 MHz		TDD	5, 10, 15, 20		
2010–2025 MHz		TDD	5, 10, 15		
1850–1910 MHz		TDD	1.4, 3, 5, 10, 15, 20		
1930–1990 MHz		TDD	1.4,3,5,10,15,20		
1910–1930 MHz		TDD	5, 10, 15, 20		
2570–2620 MHz		TDD	5, 10		EU
1880–1920 MHz		TDD	5, 10, 15, 20		
2300–2400 MHz		TDD	10, 15, 20	IMT-2000	China

Table 2.2 Frequency bands which uses FDD and TDD techniques

FDD Bands		TDD bands	
Band	**Frequencies UL/DL (MHz)**	**Band**	**Frequencies UL/DL (MHz)**
1	1920–1960/2110–2170	33, 34	1900–1920
2	1850–1910/1930–1990		2010–2025
3	1710–1785/1805–1880	35, 36	1850–1910
4	1710–1755/2110–2155		1930–1990
5	824–849/869–894	37	1910–1930
6	830–840/875–885	38	2570–2620
7	2500–2570/2620–2690	39	1880–1920
8	880–915/925–960	40	2300–2400
9	1750–1785/1845–1880		
10	1710–1770/2110–2170		
11	1428–1453/1476–1501		
12	698–716/728–746		
13	777–787/746–756		
14	788–798/758–768		
17	704–716/734–746		

cies, ambulances, radio taxi and strategic security agencies are some of the applications of the mobile radio communication service.

The transition from voice communication to wireless multimedia and web type of applications demands higher data rates. The demands on the RF spectrum by various services and applications have always far exceeded the availability of suitable spectrum. The spectrum can only be used (in a given time frame) and not consumed. It can also not be stored for future use. If not used (or not used optimally) at any given point of time, its availability and value for that point of time is lost. Also, if not used properly, it may cause harmful interference to some other system/operation, which nullifies the basic purpose/objective of the spectrum usage. Hence, effective regulation/management of this scarce natural resource at national level is essential, in the overall public/national interest.

Earlier, most of the wireless users were from government agencies, who desired compete protection from interference. Hence, exclusive assignment of frequencies for each user was essential. This method of spectrum allotment is also known as the "command and control" method of spectrum management in today's parlance. It may be added that most of the telecom operators from the private sector also want exclusive allotment of spectrum/assignment of frequencies, on the ground that they have to provide assured "Quality of Service" to their customers, which is possible if they have complete control on RF interference for their wireless based systems.

In recent years, there has been tremendous increase in usage of wireless based services by the general public for various personal purposes, e.g. cordless phones, remote controlled gadgets/toys, Wi-Fi systems, etc. [2]. Such usages are normally on shared/non-exclusive basis and most of the spectrum management organizations have earmarked some common frequency bands for such usages/applications without any specific wireless license. The technical parameters for such usage are specified to allow extensive shared public usage.

Further, to meet the growing needs of RF spectrum by the modern mobile telecom services and emerging new technologies, the spectrum managers globally are considering various alternatives to increase the efficiency of spectrum usage. These are discussed below.

New radio technologies enable more flexibility in spectrum management, unleashing innovative new products and services with appropriate spectrum policies. In the field of spectrum management the true challenges raise eminently institutional questions. The radio-frequency spectrum is essentially the raw material of present day telecommunications. This resource is increasingly in demand due to the rapid development of new radio-based services. Also, the trend of modern communications is towards increased mobility with ever increasing data rates, thus increasing the demands on the already congested spectrum.

With regard to radio network infrastructure, the scarcity of frequencies leads, at a given stage of technological development, to a natural limit on the services. Moreover, unless the market is considered as indefinitely extensible, licences for public telecom services are limited in number. These restrictions require careful planning and management for maximizing the value for all services. At a time when mobile communications are developing rapidly, the scarcity of frequencies and their management are very important issues in all countries. Tomorrow's networks will have to provide the necessary flexibility, coverage, and transparency so as to provide users total connectivity and interoperability. The deployment of such ubiquitous networks can only be feasible by capitalizing on spectrum-based resources and broadband technologies in new and innovative ways. This new environment presents technical, commercial, as well as organisational and policy challenges. The basic models of spectrum management that are at stake are the traditionally administered allocation/assignment system, the extension of property rights through primary and secondary markets, and the expanded unlicensed band approach. The unlicensed band approach has been prevalent in the earlier model/system. Also, the increased demand for and use of traditional wireless services and applications have strained the spectrum management regime. The problems are compounded by the increasingly varied demands for radio spectrum because of:

- The desire to transmit a larger quantity and a greater variety of information.
- The need for more diverse types of communications systems to meet specialized needs.
- The rapid overall changes in the technology and the marketplace.

The traditional "command and control" spectrum management techniques, based on administered allocation and assignment [4], are slowly being complemented, if not replaced, by market mechanisms such as extension of property rights of spectrum for a pre-defined limited period to private operators, through primary and secondary markets. This has raised the question whether the rule of "first come, first served" is still the best answer for optimum spectrum usage/market efficiency, whether centralized spectrum management is sustainable, and whether techniques for rationing spectrum (e.g., licences, auctions) can coexist alongside unlicensed uses. All these points raise eminently institutional questions. The progress made in electronics and microprocessors in the last ten years allows for a whole new range of products and services to be made available to the general public, such as mobile phones, satellite broadcasting of television, and radio programs with digital technology adding increasing quality and choice.

New technologies and management methods might thoroughly change the way we look at the future spectrum usage with spectrum trading and clearing (re-allocation), software defined radio: dynamic (vs. static) allocation of the spectrum, mesh networks, frequency sharing and hopping, frequency on demand, optimisation of non-permanent usages, increased system (emitter and receiver) resilience and tolerance to interference. Consequently, market mechanisms and future radio technologies challenge the premises of prevalent spectrum management principles, i.e., the long-established administered allocation and assignment process needed to prevent harmful interference and even the public domain nature of the frequency spectrum by market-based extension of property rights.

In matters of spectrum management, there is no single approach that is universally applicable, although radio spectrum is, by its nature, an international affair. There is a growing need for comparative institutional analysis. Some economies are experimenting with secondary spectrum trading, while others are using generic, technology-neutral licensing as a way of coping with the changes. Overall, there is a trend towards a greater reliance on market forces and the coexistence of multiple RF resource management methods, each adapted to a given institutional environment.

Many countries have now started to introduce some form of market-based mechanisms in managing spectrum. At one end of the scale, some European countries have retained centralized control over functions such as spectrum allocation while introducing market-based mechanisms, such as auctions, to assign spectrum. At the other end of the scale, a few countries, like Australia and Guatemala, have gone further in deregulating spectrum management by allowing a market-based allocation of spectrum use. While these market-based measures have been introduced within a system of exclusive rights, where spectrum frequencies are assigned for the exclusive use of a licensee, many countries have also allocated spectrum bands for licence-exempt use, effectively allowing more freedom for public and market players to manage the usage of spectrum among themselves.

Frequencies are used both commercially, notably for mobile communications and broadcasting, and by public sector bodies to support national defense, aviation, the emergency services and so on. As demand grows, spectrum needs to be managed to avoid harmful interference between different users. If users transmit at the same time, on the same frequency and sufficiently close to each other geographically, they will typically cause interference that might render both of their systems unusable. In some cases, "sufficiently close" might be tens or hundreds of miles apart. Even if users transmit on neighboring frequencies, they can still interfere since with practical transmitters, signals transmitted on one channel "leak" into adjacent channels, and with practical receiver's signals in adjacent channels cannot be completely removed from the wanted signal. The key purpose of spectrum management is to maximize the value that society gains from the radio spectrum by allowing as many efficient users as possible while ensuring that the interference between different users remains manageable. To fulfill this role, the spectrum manager provides each user with the right to transmit on a particular frequency over a particular area, typically in the form of a license. Clearly, the spectrum manager must ensure that the licenses that they distribute do not lead to harmful interference. In practice, this is a highly challenging task.

2.1.2 The Current Management Mechanisms

Historically, the approach [4] adopted by the spectrum managers around the world to manage the radio spectrum has been highly prescriptive. Regulators often decide on both the use of a particular band and in some cases which users are allowed to transmit in the band. Keeping a tight regulatory control over the use of the spectrum makes it easier for the regulator to ensure that harmful interference does not occur because the regulator is able to carefully model the interaction between neighboring services and tailor the license conditions appropriately. It also allows for other regulatory goals to be achieved – for example, ensuring that a service is available on a national or regional basis, or imposing coverage requirements to achieve ubiquity of services. Finally, it can result in high technical efficiency of spectrum use – that is to say in packing a large number of users into the spectrum. This is because like services in neighboring bands tend to interfere less than unlike services and so can be allocated more closely together. If the regulator collects together like services and places them adjacent to each other it can increase the capacity of the spectrum (although maximizing the capacity, or technical efficiency, is not always the same as maximizing the benefits that society can gain from the spectrum, or economic efficiency, since the spectrum can be completely used but by low value applications).

Alongside licensing certain wireless users, the spectrum manager typically exempts other usages from licensing. These exempted usages are often assigned a band of spectrum sometimes known as unlicensed spectrum, or spectrum commons. In practice, this is likely only if they transmit at a relatively low power level such that the distance over which they can operate/cause interference is small and hence the probability of there being another user within this small "coverage" area is low. Also enough spectrum bandwidth is de-licensed to allow multiple channels. Typical services that are exempt include cordless phones and wireless LANs such as Wi-Fi. It is up to the regulator to decide which equipment to exempt, what the rules for its operation should be, how much spectrum should be set aside for its operation and where in the frequency band this should be.

The current spectrum allocation process operates at both a national and international level. International coordination is essential in some cases because the zones of possible interference extend beyond national geographical boundaries and in other cases because users are inherently international, e.g. aviation, shipping, satellite systems, etc. Radio spectrum has traditionally been managed by governments. The Administrations the world over have their own National Radio Spectrum Regulatory Authority like FCC and NTIA in USA and Ofcom in the UK, etc., and at international level, it is regulated by International Telecommunication Union (ITU).

Broadly speaking, international bodies tend to set out high level guidance which national bodies adhere to in setting more detailed policy. At the highest level of management sits the International Telecommunication Union (ITU), the specialized agency of the United Nations for all global telecommunication matters. ITU's International Radio Regulations allocate the spectrum from 9 kHz to over 275 GHz to a range of different services. The Radio Regulations also set out how countries should coordinate with each other and in the case of global services, such as satellite, provides a mechanism for the assignment of rights to individual systems. The ITU conducts the key parts of its business through World Radio Conferences (WRC) which are typically held every three to four years. These are events attended by thousands of delegates from spectrum managers, technology developers and users around the globe where potential changes to the Radio Regulations are considered. In some cases, the ITU may seek international spectrum allocations for particular uses, for example in previous years it has allocated spectrum to global low Earth orbit satellite systems (of which Iridium is an example) and in its 2007 conference it has discussed whether there should be a global allocation for 4G cellular systems. Nothing in the Radio Regulations can constrain each country's freedom to manage spectrum as it wishes, as long as the impact on other countries is minimal and it is willing to accept the risk of interference from other countries, to its services which are not operating as per international regulations.

2.1.3 Shortcomings of the Current System

The current approach "works" in so much as it licenses spectrum to particular users and ensures that harmful interference is avoided. This allows a range of uses of the spectrum in a stable and predictable environment. However, it is unlikely that it achieves the full objective of a spectrum manager of maximizing the economic value derived from the spectrum. To do this, the regulator would need to make sure that spectrum was appropriately divided between all the different possible usages and users in a way which maximizes benefits to end users of spectrum based services. Since it is almost impossible to predict the value that each different service provides under any given spectrum allocation, it is difficult to see how a "command-and-control" approach to managing the radio spectrum could maximize value. However, it is possible that an extremely astute regulator might distribute spectrum in a manner that approaches this objective, at least in some bands. Widely differing valuations for the spectrum at auction, for example, the 3G auctions as opposed to spectrum auctions at 3.4 GHz suggest that the balance between different uses is incorrect. In this example, the much higher valuation of 3G suggests that there should be more spectrums made available for cellular, with perhaps less for fixed wireless or other applications.

Some new applications or technologies had great difficulty in gaining access to spectrum – for example the iBurst cellular technology or more recently mobile TV systems. While it is not certain that these would increase the value of the spectrum, their difficultly in entering the market may be a symptom of a rigid system. Some applications which have been granted spectrum free, such as aviation radar, have not modernized their radar systems for many decades despite the availability of much more efficient technologies, suggesting there are insufficient incentives for some users to optimize their use of the spectrum. The current system is also becoming increasingly difficult for the spectrum manager. They may receive frequent requests for new spectrum or to allow existing users to change applications. They may also suffer complaints of unfairness as convergence of communications services allows some users who have accessed the spectrum for free, to compete with other users who have paid for it. All of this suggests that the current approach to spectrum management, whereby the regulator selects the use of the spectrum and in some cases the user, is probably not maximizing the value of the spectrum and is becoming increasingly difficult for the regulator to administer. There is a need for a new approach to spectrum management, spectrum allocation and spectrum utilization.

2.1.4 Alternative Management Approaches

Economists have long argued that market mechanisms should be applied to RF spectrum. The essential idea here is to allow pricing mechanisms to act as an incentive for holders of spectrum to optimize their use – buying more if their business case can justify it, selling spectrum if they have excess, and adopting new technologies that can use spectrum more efficiently where economically viable. Economic theory suggests that in a market which is performing well, this will lead to a division of spectrum that maximizes economic value. Under such an approach, the regulator sets out rules that enable markets to function while ensuring that interference is controlled and then takes a back seat, leaving it to the market to determine the use and users of the spectrum. The simplest of the market instruments to adopt is probably the use of auctions as a mechanism to distribute spectrum. Auctions are now used as the preferred mechanism for assigning spectrum in many countries and they solve the most pressing problems for many of the regulators by allowing spectrum to be assigned where demand significantly exceeds supply, in a transparent and fair manner. But auctions on their own still "freeze" the assignment of spectrum. They need to be accompanied by mechanisms to trade and change the use of spectrum as market conditions change and new services become available. Technology used to lend itself to discrete allocations.

Until recently, all technologies used a relatively narrow bandwidth and assumed that they were the sole users of that frequency. For example, the GSM mobile phone system transmitted signals with a 200 kHz bandwidth, which at 900 MHz is less than a thousandth of the carrier frequency. The systems were designed assuming that there would be little interference, and where there was, it would be carefully controlled by the operator. The result of the use of these technologies has been to regulate the spectrum by frequency. That is, the spectrum is divided up into discrete parcels of frequency, for example 915–925 MHz, and assigned to a particular operator/user. That operator/user then expects that they will be given exclusive use of the band. This approach facilitates the same use of spectrum in multiple countries, often known as harmonization. By aligning the spectrum usage across multiple countries, the same technology, such as GSM, can be deployed. This brings a range of benefits including economies of scale, international roaming and reduced interference. However, it also brings some disadvantages including the need for the regulator to predict the optimum service and technology.

After over 100 years of radio spectrum management, the technology underlying the traditional approach to spectrum management is slowly changing. Two new technologies offer the potential for spectrum access under different conditions from today – these are cognitive radio and ultra-wideband. Other similar technologies might be expected to emerge in future. In addition, the ability to build multi-modal radios reduces the need for international harmonization, changing some of the drivers for conventional spectrum management.

Therefore, Flexible Spectrum Allocation [5] may need to be developed for increasing the spectrum utilization efficiency by opening the licensed spectrum band to secondary users besides primary users, while limiting the interference experienced by the primary users. Hybrid forms of the exclusive use model and the open sharing model of spectrum management models provide a view for spectrum sharing between primary & secondary users. There are two approaches for such sharing:

- *Underlay Spectrum Sharing*

  Under this approach, secondary users are allowed to operate below the noise floor described for the primary user. Ultra Wideband (UWB) technologies are particularly suited for this type of spectrum sharing because signals are spread over large swaths of spectrum and the signal strength is around the RF noise level. This allows a UWB signal to operate on occupied spectrum with a very low power output, and not cause any harmful interference to the primary user. This model relies on measuring the ambient noise and the interference caused in the operating range and maintaining it under a predefined threshold, i.e. interference measured as system temperature.

- *Overlay Spectrum Sharing*

  This approach does not impose severe restrictions on the transmission power of secondary users, but rather on when and where they may transmit. Spectrum is shared explicitly in one of three ways: opportunistic, where spectrum is used whenever the primary licensee does not use it; cooperative, where frequencies are allocated centrally based on real-time negotiation with the licensee; and mixed, where sharing is cooperative when possible and opportunistically otherwise. Dynamic Spectrum sharing comes under this model. The hybrid model seems to be more compatible with the current spectrum management policy and the underlay and overlay approaches can be employed simultaneously to further improve spectrum efficiency.

2.1.5 Dynamic Spectrum Access

Advancement in wireless technology has stimulated the demand for more radio spectrum from every corner and this demand is going to accelerate in future. Being a limited natural resource, the role of spectrum regulator becomes more difficult as to fulfill the demand of spectrum many new technology like BWA, WiMAX and 4G, etc., and at the same time to provide interference free environment for their efficient roll out. The existing practice of spectrum regulation has resulted in vastly underutilized spectrum bands not only in remote areas but also in dense urban areas. Studies have shown considerable parts of the spectrum, allocated to specific services, are practically unused for a significant period of time, meaning thereby that utilization of spectrum is relatively low not just by frequency domain but also across the spatial and temporal domain. This shows that the existing static approach to spectrum management is not efficient to cater for the present and future requirement of spectrum for new wireless applications.

Worldwide, spectrum regulators are looking for alternative methods of managing spectrum, which give a much more efficient and flexible utilization of spectrum. The solution lies with dynamic management of spectrum, which allows spectrum sharing among various wireless technologies. Spectrum sharing is not a new phenomenon. The unlicensed 2.4 GHz frequency band in which Wi-Fi and Bluetooth technology operates is one of the best examples of spectrum sharing. Such a kind of sharing is also known as horizontal sharing. This has put tremendous pressure on spectrum regulators to delicense other bands also as spectrum utilization of the licensed band is relatively low.

It is possible to open licensed bands for secondary usage with some conditions, like the operation of secondary user should not interfere with the primary user. This type of sharing is known as vertical sharing. Secondary users can access the spectrum, meant for the primary user, through Dynamic Spectrum Access (DSA) technology [5, 6], a new phenomenon in wireless communication. The key characteristic of DSA is their ability to exploit knowledge of their electromagnetic environment to adapt their operation and access to spectrum. The concept of DSA has emerged as a way to dramatically improve spectrum utilization. The basic idea behind DSA is that a device first senses the spectrum it wishes to use and characterizes the presence, if any, of primary users. Based on that information, and regulatory policies, the device identifies communication opportunities ("holes") in frequency, time, or even code and transmits using those opportunities in a manner that avoids (according to policy) creating interference that would be noticed by primary users, while attempting to limit (in accordance with spectrum etiquette) interference with other unlicensed devices operating in its vicinity.

Opportunistic spectrum access allows dramatically higher spectrum utilization. It also enables near-zero deployment time through radios that can opportunistically retarget their services to a new portion of the spectrum as needed, with obvious and significant impact on both civilian and military communications. Dynamic spectrum access is also referred to as opportunistic spectrum access, and is normally considered as part of the larger concept of cognitive radios (CR) [7].

While conceptually simple, the realization of opportunistic and dynamic spectrum access is highly challenging. There are several issues like sensing over a wide frequency band, identifying the presence of primary users and determining the nature of opportunities, coordinating the use of these opportunities with other nodes and adherence to existing regulatory policies linked with DSA, which need to be solved before implementation.

Major Benefits of Dynamic Spectrum Access

- More efficient use of spectrum
  The main reason for the development of dynamic spectrum access techniques is to provide more efficient use of spectrum, which is essentially a finite, but reusable, resource. The static model of spectrum management provides efficient use of spectrum during busy hours, but operates inefficiently at all other times. Dynamic spectrum access would enable spectrum to be used for other services during periods of low utilization, increasing efficiency. This efficiency could also be extended to creating greater device convergence, allowing multiple services to coexist in the same devices, as well as in the same spectrum.

- Faster access to spectrum
  For wireless innovation, lowering barriers to entry by making spectrum available to users who need it as quickly as possible is an important ingredient. It is difficult with static approach to quickly meet the demand for access to spectrum, particularly where spectrum has already been assigned to users. Dynamic spectrum access attempts to bypass much of the delays here by carrying out its own spectrum management.
- Greater innovation and competition
  Efficient use of and faster access to spectrum are both important to help encourage wireless innovations. This in turn results in faster time to market for innovative new wireless systems and services. An environment that supports innovation ultimately contributes to a healthy competitive environment with a greater choice of services for end users.

Potential Challenges for Dynamic Spectrum Access Interference and Quality of Service

The biggest concern over the introduction of dynamic spectrum access systems is an increased level of interference. The operation of dynamic spectrum access relies on being able to detect and avoid the transmission frequencies of other wireless systems. The ability to do this in a reliable way and in real-time can be a source of concern for existing spectrum users and spectrum managers. Interference scenarios such as the hidden node problem are important here. In some cases an increased level of interference will still enable services to function, but at a lower quality of service. Instances of interference in a dynamic spectrum environment are likely to be more difficult for spectrum managers to identify, trace and resolve than in a traditional method of spectrum management.

- The hidden node problem
  This is important in the case of a dynamic spectrum access system attempting to operate on a shared basis with other services. While the dynamic spectrum access device will "listen" to determine whether any other device is using the same frequency/ies at that time, it may not be able to identify all transmissions in its area or nearby areas due to the hidden node problem. Under certain circumstances the presence of a device will not be detected by another system seeking to use the same spectrum. The undetected device is known as a "hidden node" and it is likely to experience interference from the other system.

- Regulatory compliance
  While dynamic spectrum access systems will be designed to avoid harmful interference, there may still be a possibility that interference may occur. In such instances it would be difficult to determine the source of interference due to the way in which the spectrum is used (i.e. not in a command and control manner).

Regulatory Aspects

Nowadays the role of the spectrum regulators is becoming more difficult as they have to accommodate new services in already crowded frequency slots/bands. Regulators have to strike a balance between facilitating the rapid development of innovative services and protecting existing services and users of spectrum. There is a need to move towards more flexible and efficient systems. DSA is one of the tools which provide flexible spectrum management, which can accommodate new services with adequate protection to existing services. There are many issues which need to be addressed before implementation of DSA, like the management of Interference between primary and secondary users, which is a major concern associated with DSA. A policy should be made in such a way that secondary users should have the required freedom and at the same time the interest of the primary user should also be adequately protected. Technical neutrality and harmonization is also major concern of the regulators. Besides these, there are other issues, like the use of software in cognitive radio systems, to what extent change in parameters of secondary users can be permitted to avoid the interference, hidden node problems, etc., which would also need to be taken care of by the regulators.

Regulators have to ensure uninterrupted communication services in case of failure in the communication system (security). Other issues like international coordination balance of different types of spectrum management and keeping up the pace of technological – regulations have to be addressed.

2.1.6 Future Research

Innovative technology offers opportunities to promote spectrum access and increase its usage. This will likely happen in two ways: by enhancing the performance of traditional communications devices, or by enabling new methods to access the radio frequency spectrum. The latter seems the more relevant and challenging way to reduce artificial scarcity of spectrum, which is the crux of the debate about the reform of its management framework. Emerging technologies promise ways to make spectrum usage more intensive. However, with more systems able to access the spectrum, risks of harmful interference might be greater. In the absence of market mechanisms uncertainty about net benefits, which the technology developments discussed might offer, are great. Nevertheless, disruptive technology should not be rejected because it might bc harmful for current spectrum using services. Cost/benefits analyses that attempt to measure the impact of emerging technologies on existing services might be useful. However, this would entail a bias in favor of the latter. Therefore, future research should also evaluate net benefits in scenarios where various combinations of legacy and innovative technologies, applications and services take place. Last, but not least, such scenarios should adopt a spectrum management framework more complex and dynamic than the traditional administrative one.

2.1.7 The Indian Scenario

Presently no specific regulation has been made for the introduction of Cognitive Radio Service (CRS) in India [3]. Currently CRS technology is not fully developed [4, 5], research is progressing but reliable spectrum sensing is still a challenge. Interference between primary and secondary users can not be completely ruled out. Therefore deployment of CRS may need to be band specific at initial stage. The bands which are allocated for Aeronautical mobile services, Global Maritime Distress and Safety System (GMDSS) and safety services may not be opened for use of CRS technology as this may cause interference to these safety services. Use of CRS technologies may be permitted to those bands in which spectrum activity is low, location of base stations are known and receivers are robust against interference. Lower frequency bands, broadcasting (FM, AM and TV bands), amateur, radio paging, Public Mobile Radio Trunk Service (PMRTS) bands are some of the prominent candidates for CRS technology. Bands in which either activity is very high, e.g. GSM/CDMA band or working with very low signal strengths like Satellite services may not be opened for CRS technology till enough experience has been gained with CRS.

TV Bands

IEEE 802.22 is the standard on cognitive wireless regional area network (WRAN) which is a cognitive air interface for fixed, point-to-multipoint WRANs that operate on unused channels in the VHF/UHF TV bands. Use of IEEE 802.22 standard based WRAN [8] has already been allowed in

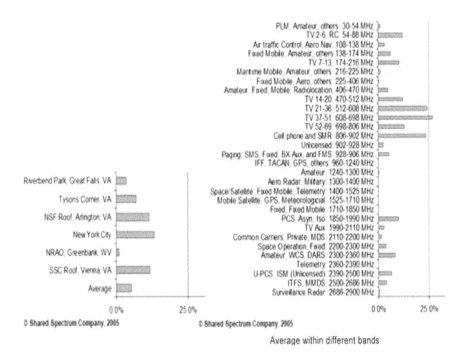

Figure 2.1 Measured spectrum occupancy at six US locations (adapted from [6])

many countries. Equipment in the market is also available. TV band can be considered for opening for WRAN in India for CRS technology.

Amateur/Radio Paging Bands

Activity in these bands is not very high. Though Indian measurements are not available, measurements taken band-wise in six US cities as in Figure 2.1 can give a fair idea of spectrum use in different bands as well in the amateur radio band.

As per NFAP 2008, those frequency bands, which are earmarked for fixed and mobile services (other than GSM/CDMA, etc.) like 235–267 MHz, 315–322 MHz, 335.4–387 MHz, 455–460 MHz (IMT application), 470–806 MHz, etc., can also be considered for utilization for CRS technology.

In addition to the above, IEEE 802.11h standard amendment defines transmission power control and a dynamic frequency selection (DFS) algorithm for wireless local area networks (WLANs) in 5 GHz band. Hence, the 5 GHz band can also be considered an opening for CRS as in Europe for WLAN.

Before deployment of CRS, extensive radio monitoring needs to be performed to gather first-hand knowledge of spectrum utilization. At present, it is recommended that lower bands starting from 200 MHz to 3 GHz may be monitored in major cities of India. It is also required to perform monitoring at three to four locations in each city on a continuous basis for at least three to four days to know the actual usage of spectrum at those locations. Based on these monitoring data, bands in which spectrum usage is low may be considered for opening for use of CRS technology.

2.2 Spectrum Efficiency and Regulatory Aspects for Wireless/Mobile Systems Using Cognitive Radio

In this section, we will discuss the traditional notion regarding the spectrum efficiency and the definition of modern definition of spectrum efficiency in context to Cognitive radio paradigm. This paradigm calls for spectrum reuse and sharing of the legacy bands. This will affect the regulatory policies and will demand formulation of new policy framework.

2.2.1 Introduction to Cognitive Radio

We are moving at a fast pace towards a "Mobile Wireless Society". First generation mobile radio systems were analog voice telephone systems. They were first deployed in the early 1980s, and provided reasonably good quality and capacity. In addition to voice systems, second generation mobile radio systems, which use digital technology in contrast to the first generation systems, provided low rate data services and other auxiliary services. Second generation systems are very successful worldwide in providing service to users. GSM is the mobile radio standard with today's highest penetration worldwide. Recent advances have enabled wireless handsets with data download capability of 1-10Mbps. In the next few years, this will be extended to roughly 100Mbps and then on to over 1Gbps in the next decade. This ability to handle data at very high speed will enable consumers to easily handle high quality audio, video, and high resolution images. Advanced CDMA air interface technologies offer spectral efficiencies and with OFDM technology and MIMO techniques, it is anticipated that much higher spectral efficiencies will be achieved. The future of telecommunications is anticipated to be an evolution and convergence of mobile communication systems with IP networks, leading to the availability of a great variety of innovative services over a multitude of Radio Access Technologies (RATs). To achieve this vision, it is mandatory to embrace the requirements for the support of heterogeneity in wireless access technologies, comprising different services, mobility patterns, device capabilities, and so on.

2.2.2 Ultra-Wideband (UWB)

The future UWB is a technology for transmitting information spread over a large bandwidth (>500 MHz) that should, in theory and under the right circumstances, be able to share spectrum with other users. The UWB can use the frequency 3.1–10.6 GHz. This is intended to provide an efficient use of scarce radio bandwidth while enabling both high data rate *personal-area network* (PAN) wireless connectivity and longer-range, low data rate applications as well as radar and imaging systems. Ultra Wideband was traditionally accepted, the ITU-R now define UWB in terms of a transmission from an antenna for which the emitted signal bandwidth exceeds the lesser of 500 MHz or 20% of the center frequency. A significant difference between traditional radio transmissions and UWB radio transmissions is that traditional transmissions transmit information by varying the power/frequency and/or phase of a sinusoidal wave. UWB transmissions can transmit information by generating radio energy at specific time instants and occupying large bandwidth thus enabling a pulse-position or time-modulation. One of the valuable aspects of UWB radio technology is the ability for a UWB radio system to determine "time of flight" of the direct path of the radio transmission between the transmitter and receiver to a high resolution. Due to the extremely low emission levels currently allowed by regulatory agencies, UWB systems tend to be short-range and indoors. However, due to the short duration of the UWB pulses, it is easier to engineer extremely high data rates, and data rate can be readily traded for range by simply aggregating pulse energy per data bit using either simple integration or by coding techniques. Conventional OFDM technology can also be used subject to the minimum bandwidth requirement of the regulations. UWB differs substantially from conventional narrowband radio frequency (RF) and spread spectrum technologies (SS), such as Bluetooth Technology and 802.11a/g. UWB uses an extremely wide band of RF spectrum to transmit data. In so doing, UWB is able to transmit more data in a given period of time than the more traditional technologies. Bluetooth Technology, 802.11a/g Wi-Fi, cordless phones, and numerous other devices are relegated to the unlicensed frequency bands that are provided at different frequency bands. Each radio channel is constrained to occupy only a narrow band of frequencies, relative to what is allowed for

UWB. UWB is a unique and new usage of a recently legalized frequency spectrum. With the characteristics of low power, low cost, and very high data rates at limited range, UWB is positioned to address the market for a high-speed WPAN. UWB technology also allows spectrum reuse. A cluster of devices in proximity (for example, an entertainment system in a living area) can communicate on the same channel as another cluster of devices in another room (for example, a gaming system in a bedroom). UWB-based WPANs have such a short range that nearby clusters can use the same channel without causing interference. An 802.11g WLAN solution, however, would quickly use up the available data bandwidth in a single device cluster, and that radio channel would be unavailable for reuse anywhere else in the home. Because of UWB technology's limited range, 802.11 WLAN solutions are an excellent complement to a WPAN, serving as a backbone for Data transmission between home clusters. UWB technology has distinct advantages such as:

- Enabling high-speed wireless universal serial bus (WUSB) Connectivity for PCs and PC peripherals, including printers, scanners, and external storage devices; and
- Replacing cables in next-generation Bluetooth Technology; and devices, such as 3G cell phones, as well as IP/UPnP-based

2.2.3 Multiband, Multimode, Multimedia Wireless Devices

Recently AT&T announced the first "four-mode handset", which can operate in TDMA mode on 850 or 1900 MHz, in first generation Analog Mobile Phone System (AMPS) mode on 850 MHz, and in Cellular Digital Packet Data (CDPD) mode. This is just the beginning of the multiband, multimode, multimedia wireless explosion. In the not-too-distant future, software-radio based Personal Digital Assistants (PDAs) could access satellite mobile services, cordless telephone, RF LAN, GSM, and 3G W-CDMA. Such a device could affordably operate in octave bands from 0.4 to 0.96 GHz, (skip the air navigation and GPS band from 0.96 to 1.2 GHz), 1.3 to 2.5 GHz, and from 2.5 to 5.9 GHz. This wideband radio technology will be affordable first for infrastructure, next for mobile vehicular radios and later for handsets and PDAs. Such software radio technology expands opportunities for the dynamic sharing of spectrum. Cognitive radio would make such sharing practical.

2.2.4 Software Defined Radio (SDR)

The proliferation of wireless communications services has caused a con-current increase in the demand for and congestion of radio frequency (RF) spectrum. This congestion has put a premium on the cost of spectrum and has created a battle between the public, private, and military sectors over frequency ownership. Studies have shown, however, that spectral utilization is relatively low when examined not just by frequency domain, but across the spatial and temporal domains. Thus, an intelligent device aware of its surroundings and able to adapt to the existing RF environment in consider-ation of all three domains, may be able to utilize spectrum more efficiently by dynamically sharing spectral resources. This type of real-time spectrum monitoring and transmission method alteration requires that certain technical and operational situations exist. The device must be capable of sensing its op-erating environment. A SDR is a transmitter in which operating parameters, including transmission frequency, modulation type, and maximum radiated or conducted output power can be altered without making any hardware changes. The classical method of demodulating high frequency signals is to shift the signal's frequency down in one or two steps before feeding it to a demodulator. The frequency shifting is necessary because the demodulation is more stable and easier to implement at lower frequencies. The demodulator block is designed to perform a fixed set of services (e.g., modulation and demodulation, encryption and decryption) chosen at design time. Advances in computer design and digital signal processing (DSP) make it possible to replace the classic demodulator with a component that dynamically sup-ports multiple systems, protocols, and interfaces. The signal in this case is sampled by a high-speed analog-to-digital converter (ADC) and all the processing required to extract the useful signal is performed by a computer. This new approach enables an unprecedented flexibility in modulation and encryption capabilities because the digital signal processing module can be reprogrammed. Furthermore, the separation of the hardware platform from the functionality allows the emergence of standards for the hardware plat-form which, in turn, offers tremendous benefits for the development and deployment of SDR devices. Cognitive radio may help improve spectrum management by moving it from the sclerotic framework of regulations to the flexible realm of networks and devices, thereby enabling dynamic spectrum sharing and improving spectrum utilization.

2.2.5 Cognitive Radio

The concept of Cognitive Radio (CR) [7] emerged as an extension of SDR technology. Definitions vary, but most experts agree that a CR device should have the following characteristics:

- aware of its environment,
- capable of altering its physical behavior to adapt to its current environment,
- learns from previous experiences, and
- deals with situations unknown at the time of the radio's design.

In practice, we consider multiple levels of cognition, the simplest being a pre-programmed device that has no model-based reasoning, but can sense its operating environment and make some decisions about which built-in capabilities to use at a given time. An example of such a device is a multi-band, multi-protocol cellular telephone. Higher levels of cognition – which are not available today – would allow a cognitive radio to negotiate the parameters of the communication with other radios or base stations.

The limited available spectrum and the inefficiency in spectrum usage necessitate a new communication paradigm to exploit the existing wireless spectrum. This new networking is known as Next generation (xG) network. The xG network will provide high bandwidth to users by using dynamic spectrum access techniques. The inefficient usage of the existing spectrum can be improved through opportunistic access to the licensed bands without interfering with the existing users. The key factor of xG networks is the cognitive radio. Cognitive radio techniques provide the capability to use or share the spectrum in an opportunistic manner. Dynamic spectrum access techniques allow the cognitive radio to operate in the best available channel. More specifically, the cognitive radio technology will enable the users to:

- determine which portions of the spectrum is available and detect the presence of licensed users when a user operates in a licensed band (spectrum sensing),
- select the best available channel (spectrum management),
- coordinate access to this channel with other users (spectrum sharing), and
- vacate the channel when a licensed user is detected (spectrum mobility).

The xG network can operate in both licensed and unlicensed bands. The functionalities required vary according to whether spectrum is licensed or unlicensed.

2.3 Spectrum and Regulatory Aspects: State-of-the-Art: Journey So Far

It needs to be discussed about the challenges in spectrum management and regulation since the access technology needs to be defined anew for future telecommunication networks.

2.3.1 Radio Frequency Spectrum

The radio spectrum – the segment of the electromagnetic continuum containing waves in the radio-frequency range – accommodates countless communications devices today. The tremendous growth in the wireless based systems and the evolution of the radio communication technologies at a much faster rate have put great pressure on radio frequency spectrum management. This pressure has put a premium on the cost of spectrum and has created a battle between the public, private, and military sectors over frequency ownership. At present, the radio spectrum managers allocate the radio spectrum with frequency of varying widths. One band covers AM radio, another VHF television; still others cell phones, and so on. In fact, in some locations or at some times of day, 70% of the allocated spectrum may be sitting idle, even though it is officially spoken for. Moreover, there are many areas of the radio spectrum which are not fully utilized in different geographical areas of the country. The problem is not the lack of radio spectrum but the way that spectrum is being used. Studies have shown that spectral utilization is relatively low when examined not just by frequency domain, but across the spatial and temporal domains. We need a device that must be capable of sensing its operating environment, determining the best transmission method, and reconfiguring itself to the ideal state. There are three spectrum management models exist at present in the most of the countries – command and control, exclusive use, and unlicensed use, often referred to as "spectrum commons".

The increased demand for and use of traditional wireless services and applications have strained the spectrum management regime. The use of unlicensed spectrum bands has exploded and will soon cause severe congestion problems and the command and control governance of public safety spectrum has been questioned because of low, non-peak time utilization rates. Unlicensed spectrum use provides no interference protection rights, and as the band becomes increasingly congested, quality of service will fall. Under the command and control model, entities that require dedicated, 100% exclusive use spectrum, i.e., public safety agencies must have immediate access to all of their assigned frequencies in an incident response situation. At all other times, however, their spectrum is not being fully utilized and is therefore not fully efficient. The exclusive use model affords spectrum licensees the flexibility to use their spectrum as the market demands, provided that use does not interfere with other wireless services. Spectrum access negotiation, however, is limited to the speed with which that two or more parties can agree on terms and conditions.

2.3.2 Regulatory Bodies

Administrations the world over have their own National Radio Spectrum Regulatory Authority. For example, the US operate in a bifurcated spectrum management system where the Federal Communications Commission (FCC) governs commercial wireless communications users and state and local government spectrum users, while the National Telecommunications and Information Administration (NTIA) regulates all Federal Government spectrum users. Additionally, the US conforms to any applicable International Telecommunication Union (ITU) regulations as set forth comments under any set timeframe, often making the process quite lengthy. The FCC is charged with acting in the public's "best interest", a somewhat nebulous statutory term whose definition is often considerably debated. In India the Wireless Planning & Coordination (WPC) Wing of the Ministry of Communications & IT, created in 1952, is the National Radio Regulatory Authority responsible for Frequency Spectrum Management, including licensing and caters for the needs of all wireless users (Government and Private) in the country. It exercises the statutory functions of the Central Government and issues licenses to establish, maintain and operate wireless stations. The WPC Wing performs its functions through various divisions/groups. The Standing Advisory Committee on Radio Frequency Allocation (SACFA) makes the recommendations on major frequency allocation issues, formulation of the frequency allocation plan, making recommendations on the various issues related to International Telecom Union (ITU), to sort out problems referred to the committee by various wireless users, citing clearance of all wireless installations in the country, etc. In pursuance of the New Telecom Policy, 1999, the National Frequency Allocation Plan-2000 (NFAP-2000) evolved and was made effective from January 2000, which formed the basis for development, manufacturing and spectrum utilization activities in the country. At the time of formulation of NFAP-2000, it was recognized that there will be a need to review the NFAP generally every two years in line with the Radio Regulations of the International Telecommunication Union (ITU) in order to cater to newly emerging technologies as well as to ensure equitable and optimum utilization of the scarce limited natural resource of radio frequency spectrum. Accordingly, NFAP-2000 has been revised and new National Frequency Allocation Plan-2002 (NFAP-2002) and NFAP-05 have been evolved within the overall framework of the ITU, taking into account spectrum requirements of the government as well as the private sector.

The ITU's radio communication sector coordinates spectrum use on an international level, seeking to globally harmonize RF spectrum bands and to reduce harmful interference between countries to improve the use of RF services. Any cognitive device would be subject to the rules and regulations set forth by the respective regulatory bodies, under every operating condition of the device. As the device reconfigures itself, it must be aware of, or able to download software defining the rules associated with its operating status and geographic location. In a Spectrum Management agency, three spectrum management models may exist – command and control, exclusive use, and unlicensed use, often referred to as "spectrum commons". Rules governing spectrum use under these models are as widely varied as the motivations for providing the services that they regulate. The command and control model has been the traditional spectrum management model. The principle behind it is that RF spectrum is divided into frequency bands, in which specific channels are licensed to specific users for specific services. These frequency bands are subject to explicit usage rules governing the designated RF service or transmission type. Under the exclusive use model, a licensee is authorized to use a specific frequency band or channel for whatever service or purpose they desire. These rights are subject to general emission rules that are designed not to interfere with neighboring spectrum users. In return for not interfering with other users, licensees are afforded the freedom to use their spectrum however they choose, much like property rights. RF devices governed by the tenets of the spectrum commons, which typically refers to unlicensed frequency bands, operate on a first-come, first-serve basis. Devices transmitting in this band are subject to certain emission rules, but are not guaranteed any interference protection rights or the exclusive use of dedicated channels.

The Spectrum Management Vision calls for:

- Technology Neutrality: the freedom of an operator to use whatever radio standard he wants in a certain spectrum. Technology neutrality in the field of radio telecommunication therefore means in fact "*Standards Neutrality*".

- Technology Harmonization: as a process to increase commonalities and to decrease differences with the aim to improve interoperability and compatibility (between standards). A primary goal of reconfigurability is to realize technology harmonization. From the spectrum usage viewpoint, *harmonized spectrum* is a prerequisite for economics of scale and eases global roaming.
- Harmonization versus Technology Neutrality: harmonization and technology neutrality can be considered as antithetical tendencies. Whilst harmonization aims at minimization or even avoidance of differences between standards, technology neutrality aims to permit different standards and their usage, and to leave it to the market preferences and other influences whether alignment will occur or not. Harmonization aims at maximum interoperability and compatibility for mobile communication, whilst technology neutrality consciously risks the usage of different incompatible standards resulting in non-interoperability and missing roaming capabilities between different systems in adjacent or even same areas and/or spectrum bands.

With the existing radio regulatory regime, whenever radio services for new wireless applications evolve, the status of spectrum allocation and licensing has to be changed; new radio spectrum has to be made available. However, changing the status of licensed radio spectrum may be slow as it has to take a concerted effort among government regulatory agencies, technology developers and service providers to achieve efficient and timely deployment. This is one of the reasons why, paradoxically, 90–95% of the licensed radio spectrum is not in use at any location at any given time. The existing radio regulatory regime is simply too complex to handle the increasingly dynamic nature of emerging wireless applications. As a result, we waste precious spectrum. On the contrary, the commercial success of wireless applications in the unlicensed bands, and the many radio systems utilizing this fraction of the radio spectrum, indicate that it may be helpful to change our so much established and manifested radio regulatory regime, towards a more flexible, open spectrum access. We may just want to let radio systems coordinate their usage of radio spectrum themselves, without involving regulation. Self-organizing radio systems would then autonomously regulate in a technology-based approach, here the machines would make the decisions and not humans. It can only be imagined how our economies would gain from such a flexible, technology-based approach. If successful, this approach would support new emerging wireless applications, at the same time allowing existing incumbent radio services to continue operating without significant quality-of-service (QoS) degradation. Such an open approach would potentially increase the usage of our radio spectrum significantly. There are two main challenges that exist with the proposed technology based approach for regulation:

- Protection of the incumbents: The first challenge for cognitive radio is to protect the operation of incumbent, licensed radio services. The additional opportunistic usage of radio spectrum should not harm the operation of already established radio services.
- Coexistence: The other challenge for cognitive radio is to overcome the problem of coexistence and potential free riders in an openly accessible spectrum. Apparently, when radio systems share spectrum, it is difficult to guarantee that spectrum usage is not monopolized by some radio systems, while at the same time other radio systems would not be able to operate.

It is true that advanced spectrum management will be a hot topic in future research on wireless networks as Unlicensed (ISM, WLAN) as well as secondary (UWB) spectrum usage are already under way and first spectrum sharing strategies (e.g. spectrum pooling) are under investigation. Advanced spectrum management calls for new developments in networking and terminal. It is also evident that:

- Some frequency bands in the spectrum are largely unoccupied most of the time.
- Some other frequency bands are only partially occupied.
- The remaining frequency bands are heavily used.

In its recent Communication of 2007, the European Commission argues that "the deployment of innovative wireless services and technologies is increasingly hampered by the reservation of certain spectrum bands for narrowly defined services coupled with rigid usage conditions that are unduly constraining spectrum use". In fact, measurements of spectrum usage have demonstrated that, even in urban areas, the allocated radio frequencies are often under used. For instance, during some spectrum occupancy measurements to study for determination of the spectrum occupancy in each band (30–3,000 MHz), it was demonstrated that the average occupancy over all of the locations and the maximum total spectrum occupancy were not more than 15–18%. The underutilization of the electromagnetic spectrum leads us to think in terms of spectrum holes, that is to say a spectrum hole is a band of frequencies assigned to a primary user, but, at a particular time and specific geographic location, the band is not being utilized by that user. Spectrum utilization can be improved significantly by making it possible for a secondary user (who is not being serviced) to access a spectrum hole unoccupied by the primary user at the right location and the time in question.

2.3.3 State-of-Art of Cognitive Radio (CR) and CR-vs-Spectrum

Software Defined Radio (SDR) and CR technology advancements have the potential to alleviate the limitations in the frequency, spatial, and temporal domains and provide for real-time spectrum access negotiation and transactions, thus facilitating dynamic spectrum sharing. The idea of CR devices developed naturally as an extension of SDR technology. The general-purpose computing capability that is essential to the SDR architecture can be harnessed to provide the functionality needed for "cognition", as defined at the beginning of this paper. As this computing power increases, more sophisticated cognition algorithms can be implemented. Equally important for the cognition process is environment sensing. While there are many ways of sensing the external world, being aware of the surrounding radio environment calls for smart antennas, which are capable of sensing across wide bands of the radio spectrum. These types of antennas are being developed in an effort to improve the frequency agility of SDR. Several obstacles must be overcome before cognitive radio implementation can become practical. In the underlay sharing model, the radio needs to be aware of the ambient RF noise level as a function of both space and time. With today's technology it is fairly difficult to measure the interference temperature threshold given that in real-life situations, directional and omni-directional RF sources can coexist. The overlay spectrum sharing mode calls for higher computing power and more sophisticated radio reconfigurability than current technology offers. Efficient spectrum negotiation algorithms have yet to be developed. Security is also a major concern. In an environment where radios can decide how spectral resources are used, proper authentication is vital. Rogue players have the potential to cause major communication disruption. Mechanisms must be developed to identify and incapacitate these devices. One of the defining characteristics of SDR and cognitive radio is the ability to gain or modify functionality through over-the-air software downloads. As the number of wireless technologies, products, and users increases, many applications need more bandwidth. Then cellular and industrial, scientific, and medical (referred to as ISM) bands are presently over-utilized. In contrast, television and land mobile radio bands are most often under-utilized; primarily during periods of off-peak use.The idea of cognitive radio was first presented officially in an article by Mitola [7]. It was a novel approach in wireless communications that Mitola later described as the point in which wireless personal digital assistant (PDAs) and the related networks are sufficiently computationally intelligent about radio resources and related computer-to-computer communications to detect user communications needs as a function of use context, and to provide radio resources and wireless services most appropriate to those needs. Cognitive radio should have the following characteristics:

- Aware of its environment.
- Capable of altering its physical behavior to adapt to its current environment.
- Learns from previous experiences.
- Deals with situations unknown at the time of the radio's design.

The term "cognitive radio" can formally be defined as follows. A "cognitive radio" is a radio that can change its transmitter parameters based on interaction with the environment in which it operates. From this definition, two main characteristics of the cognitive radio can be defined.

Cognitive Capability

Cognitive capability refers to the ability of the radio technology to capture or sense the information from its radio environment. This capability cannot simply be realized by monitoring the power in some frequency band of interest but more sophisticated techniques are required in order to capture the temporal and spatial variations in the radio environment and avoid interference to other users. Through this capability, the portions of the spectrum that are unused at a specific time or location can be identified. Consequently, the best spectrum and appropriate operating parameters can be selected. The three main steps of the cognitive capability are spectrum sensing, spectrum analysis, and spectrum decision.

- Spectrum sensing: A cognitive radio monitors the available spectrum bands, captures their information, and then detects the spectrum holes.
- Spectrum analysis: The characteristics of the spectrum holes that are detected through spectrum sensing are estimated.
- Spectrum decision: A cognitive radio determines the data rate, the transmission mode, and the bandwidth of the transmission. Then, the appropriate spectrum band is chosen according to the spectrum characteristics and user requirements.

Once the operating spectrum band is determined, the communication can be performed over this spectrum band. However, since the radio environment changes over time and space, the cognitive radio should keep track of the changes of the radio environment. If the current spectrum band in use becomes unavailable, the spectrum mobility function is performed to provide a seamless transmission. Any environmental change during the transmission such as primary user appearance, user movement, or traffic variation can trigger this adjustment.

Reconfigurability

The cognitive capability provides spectrum awareness whereas reconfigurability enables the radio to be dynamically programmed according to the radio environment. In other words, reconfigurability is the capability of adjusting operating parameters for the transmission on the fly without any modifications on the hardware components. This capability enables the cognitive radio to adapt easily to the dynamic radio environment. There are several reconfigurable parameters that can be incorporated into the cognitive radio as explained below:

- Operating frequency: A cognitive radio is capable of changing the operating frequency. Based on the information about the radio environment, the most suitable operating frequency can be determined and the communication can be dynamically performed on this appropriate operating frequency.
- Modulation: A cognitive radio should reconfigure the modulation scheme adaptive to the user requirements and channel conditions. For example, in the case of delay sensitive applications, the data rate is more important than the error rate. Thus, the modulation scheme that enables the higher spectral efficiency should be selected. Conversely, the loss-sensitive applications focus on the error rate, which necessitate modulation schemes with low bit error rate.
- Transmission power: Transmission power can be reconfigured within the power constraints. Power control enables dynamic transmission power configuration within the permissible power limit. If higher power operation is not necessary, the cognitive radio reduces the transmitter power to a lower level to allow more users to share the spectrum and to decrease the interference.
- Communication technology: A cognitive radio can also be used to provide interoperability among different communication systems. The

transmission parameters of a cognitive radio can be reconfigured not only at the beginning of a transmission but also during the transmission. According to the spectrum characteristics, these parameters can be reconfigured such that the cognitive radio is switched to a different spectrum band, the transmitter and receiver parameters are reconfigured and the appropriate communication protocol parameters and modulation schemes are used.

The ultimate objective of cognitive radio is to obtain the best available spectrum through cognitive capability and reconfigurability. Since most of the spectrum is already assigned, the most important challenge is to share the licensed spectrum without interfering with the transmission of other licensed users. The cognitive radio enables the usage of temporally unused spectrum, which is referred to as spectrum hole or white space. If this band is further used by a licensed user, the cognitive radio moves to another spectrum hole or stays in the same band, altering its transmission power level or modulation scheme to avoid interference.

The long-term vision of cognitive radio technology is one in which handsets would automatically make use of underutilized spectrum across a broad frequency range, allowing the high bandwidth requirements of the future set out in our vision to be realized. An RF cognitive radio card can therefore turn a cell phone into a WLAN, a laptop into a cell phone or a cordless telephone into a picocell "tower". From such a picocell, a home computer fitted with a cognitive radio control system could rent airtime to passers-by, billing for secure wireless voice or data through the associated Internet service provider.

2.3.4 Application of Cognitive Radio on Licensed Band

Use of cognitive radio relies on the concept of the possibility of permitting unlicensed users to temporarily "borrow" spectrum from licensed holders as long as no undue interference is seen by the primary user. Basic cognitive radio techniques, such as dynamic frequency selection and transmit power control, already exist in many unlicensed devices. However, to reach the full promise of cognitive radios, many significant design challenges lie ahead. Before beginning operations, cognitive radios must obtain an estimate of the power spectral density of the radio spectrum to determine which frequencies are used and which frequencies are unused. In order to accurately measure the spectrum, a highly sensitive radio will be required to measure signals at their cell edge. Consider the example of digital TV which lies at the edges; the received signal will be just barely above the sensitivity of the receiver. For a cognitive radio to be able to detect this signal, it needs to have a radio that is considerably more sensitive. If the cognitive radio is not capable of detecting the digital TV signal, then it will incorrectly determine that the spectrum is unused; thereby leading to potential interference if this radio spectrum is used, i.e., the signal transmitted by the cognitive radio will interfere with the signal the digital TV is trying to decode. This situation is often referred to as the "hidden node problem".

Further, when the spectrum is occupied, the cognitive radio must also be able to estimate the "interference temperature" that the primary user can tolerate, i.e. the transmit power level that a cognitive device can utilize without raising the noise floor of the primary user's device beyond a specified amount. In many cases, this specified amount is in the order of 0.5 to 1.0 dB, but will depend heavily on the link margin available at the primary user's receiver. The interference temperature can be determined with at least two pieces of information: an estimate of the signal bandwidth used by the primary user, and the distance between the cognitive radio and the primary user's device. The signal bandwidth can be used to determine the noise floor for the primary user's device, while the distance can be used to determine the received signal strength seen by the primary user's device as a function of transmit power used at the cognitive radio. Assuming that the noise floor at the primary user's receiver is allowed to rise by a pre-specified amount, it is easy to calculate the maximum allowed transmit power for the cognitive radio. Of course, this analysis is very simplistic and can be refined even further if the cognitive radio can blindly classify the type of signal and corresponding data rates used by the primary user. This extra knowledge would determine the exact sensitivity requirements for the primary user's device. Since the spectrum use is constantly changing, the estimate will need to be continuously updated in order to ensure that the primary users are always being protected from interference. Additionally, a simplistic view of the radio transmission path is assumed. In a real-world case, the propagation path from the cognitive radio transmitter to the primary user's receiver might be quite complicated. For example, obstacles between the two could substantially attenuate the signal, meaning that the cognitive radio could transmit at a much higher power level than would normally be assumed. However, the alternate possibility, that reflections of the transmitted signal could enhance interference at the primary receiver, is also possible. Consequently, lack of information regarding the transmission paths between the various transmitters and receivers in a cognitive radio network would create a very serious design challenge. There is a significant challenge also lies in the design of the receiver of the cognitive radio. Emissions from primary users will also result in interference for the cognitive radio. Since emissions from primary users are uncontrollable, the cognitive radios must use advanced radio designs, such as a multi-user detector, to deal with the interference and ensure that the desired signals can be reliably decoded. Finally, to be able to take advantage of the ever changing spectrum conditions, the cognitive radios need to be adaptable in the spectrum, power levels, modulation schemes, and protocols that they use. This is especially challenging for mobile radio environments in which channel conditions can change rapidly when vehicular speeds are involved.

2.3.5 Cognitive Radio on Unlicensed Band

Open spectrum policy that began in the industrial scientific and medical (ISM) band has caused an impressive variety of important technologies and innovative uses. However, due to the interference among multiple heterogeneous networks, the spectrum efficiency of ISM band is decreasing. Ultimately, the capacity of open spectrum access, and the quality of service they can offer, depends on the degree to which a radio can be designed to allocate spectrum efficiently. Cognitive radio can be designed for operation on unlicensed bands such that the efficiency is improved in this portion of the spectrum. Since there are no license holders, all users have the same right to access the spectrum bands. Multiple users coexist in the same area and communicate using the same portion of the spectrum. Intelligent spectrum sharing algorithms can improve the efficiency of spectrum usage and support high QoS. However, since all users have the same right to access the spectrum, users should compete with each other for the same unlicensed band. Thus, sophisticated spectrum sharing methods among users are required. If multiple network operators reside in the same unlicensed band, fair spectrum sharing among these networks is also required. User priority is another aspect of spectrum sharing systems that must be considered. The use of genetic algorithms has been proposed for this purpose. Of course, many aspects of the use of cognitive radio in unlicensed band need additional research.

2.3.6 Cognitive Radio and Spectrum Management

Cognitive radio technology can also simplify regulatory policy. Since users can share spectrum, the demand for special use licenses should be reduced. The cognitive radio technologies could meet ever increasing requirements for spectral efficiency, effective etiquette, and resistance to interference. It is noteworthy that the cognitive radio network concept has the potential to explicitly protect the spectrum rights of incumbent license holders. License holders can be guaranteed immediate and top priority access to their spectrum and all spectrums made available for the network use on a temporary basis could be rapidly recovered by the primary license holder if needed. A good example of this is the potential for shared use of spectrum reserved today for military use. In most parts of the country, this spectrum goes unused most of the time, making it ideal for shared use spectrum. However, in the event of an emergency, a cognitive radio network based system could easily evacuate the full band on a nationwide level for immediate and exclusive military use. This ability is also critical to cellular operators, where the large additional spectrum pool, dramatically reduced costs, and ability to guarantee service through use of incumbent licensed spectrum makes the cognitive radio based network very attractive. With increased focus over the past few years on system security and survivability, it is also important to note that distributed intelligent systems, such as cognitive radio, offer benefit in the event of attacks or natural disasters. While a conventional cellular system depends heavily on centralized control, a cognitive system is capable of establishing communications even if some network elements are out of order. Since spectrum is shared, spectrum can continue to be used even if one entity's (or network operators) system has failed.

In spite of the technical challenge, it is clear that existing approaches to spectrum management, which date to the very early days of radio technology, will soon be insufficient to meet the demands of modern wireless communications. Cognitive radio offers hope to meet this demand with a system that is compatible with existing deployed wireless systems, stimulates new innovation, reduces regulatory burden, encourages market competition, preserves the rights of incumbent spectrum license holders, and benefits the populace overall.

Current Spectrum Management Research

CORVUS: The Cognitive Radio approach for usage of Virtual Unlicensed Spectrum (CORVUS) system is presented to exploit unoccupied licensed bands. In CORVUS, based on the local spectrum sensing, the primary user detection and the spectrum allocation are performed in a coordinated manner. This cooperative effort greatly increases the system's ability in identifying and avoiding primary users. In CORVUS, a group of users form a secondary user group (SUG) to coordinate their communication. Each member of this group senses the spectrum pool, which is divided into sub-channels. A universal control channel is used by all groups for coordination and separate group control channels are used by the members of a group to exchange sensing information and establish secondary user links. The performance of the physical and link layers are evaluated through the CORVUS test bed. Moreover, recently, a reliable link maintenance protocol is proposed within CORVUS to maintain the quality of secondary user communication.

The OCRA network: The OFDM-based cognitive radio (OCRA) network considers all possible deployment scenarios over the heterogeneous xG network environment and develops cross-layer operations for the OFDM based dynamic spectrum access. For the spectrum decision and the spectrum handoff, OCRA network provides a novel concept of an OFDM-based spectrum management over the heterogeneous spectrum environment. Based on this physical layer (PHY) structure, a dual-mode spectrum sharing framework is proposed, which enables access to existing networks as well as coordination between xG users. Furthermore, a new routing paradigm that considers joint re-routing and spectrum handoff is proposed. Moreover, OCRA network introduces multi-spectrum transport techniques to exploit the available but non-contiguous wireless spectrum for high quality communication. The OCRA testbed is based on IEEE 802.11a/g technology, which exploits the OFDM technology.

IEEE 802.22: This is the first worldwide standard based on cognitive radio technology and is now in the process of standardization. This project, formally called the standard for wireless regional area networks (WRAN), focuses on constructing fixed point-to-multipoint WRAN that will utilize UHF/VHF TV bands between 54 and 862 MHz. Specific TV channels as well as guard bands will be used for communication in IEEE 802.22. The IEEE 802.22 system specifies a fixed point-to-multipoint wireless air interface whereby a base-station manages its own cell and all associated users, which are denoted as consumer premise equipments (CPEs). IEEE 802.22 base-station manages a unique feature of distributed sensing. This is needed to ensure proper incumbent protection and is managed by the base-station, which instructs the various CPEs to perform distributed measurement activities. A distinctive feature of IEEE 802.22 WRAN as compared to the existing IEEE 802 standards is the base-station coverage range, which can go up to 100 km if the power is not an issue. Current specified coverage range is 33 km at 4 W CPE effective isotropic radiated powers (EIRP).

Spectrum Management Research in the US

A considerable amount of research has been done in the US on spectrum management technologies including cognitive radio. Some of the research on cognitive radio and related topics has been fostered by the Federal Communications Commission concepts on cognitive radio and interference temperature as described in the FCC Spectrum Policy Task Force Report. This chapter noted that: "Preliminary data and general observations indicate that portions of the radio spectrum are not in use for significant periods of time". The FCC report notes, however, that more information is needed in order to quantify and characterize spectrum usage more accurately so that the FCC can adopt spectrum policies that take advantage of the "holes" in the spectrum usage. Subsequent to the issuance of the Spectrum Policy Task Force Report, the FCC has issued other related documents including:

- Facilitating Opportunities for Flexible, Efficient, and Reliable Spectrum Use Employing Cognitive Radio Technologies;
- Establishment of an Interference Temperature Metric to Quantify and Manage Interference and to Expand Available Unlicensed Operation in Certain Fixed, Mobile and Satellite Frequency Bands;
- Unlicensed Operation in the TV Bands; and
- New paradigm for spectrum management – cognitive radio, policy-based radio, reconfigurable radios and networks.

The Defense Advanced Research Projects Agency (DARPA) is developing a new generation of spectrum access technology under the Next Generation (**xG**) communications program. Although this program is oriented toward military communications, it is applicable to advanced spectrum management for any band and any communications services. This multiusage is analogous to the development of data communications protocols which were originated by DARPA and used in the DARPA-net, but which have involved to what is now known as the World Wide Web. The motivation for this new spectrum access technology is the same for the commercial communications community as it is for the military communications community:

The National Science Foundation (NSF) has a research program entitled "Programmable Wireless Networking" (NeTS-ProWiN). This NSF research program addresses issues that result from the fact that wireless systems today are characterized by wasteful static spectrum allocations, fixed radio functions, and limited network and systems coordination. This has led to a proliferation of standards that provide similar functions – wireless LAN standards (e.g., Wi-Fi/802.11, Bluetooth) and cellular standards (e.g., 3G, 4G,

CDMA, and GSM) – which in turn has encouraged stovepipe architectures and services and has discouraged innovation and growth. Emerging programmable wireless systems can overcome these constraints as well as address urgent issues such as the increasing interference in unlicensed frequency bands and low overall spectrum utilization. The research sponsored by NSF under the Programmable Wireless Networking Program is addressing these issues by supporting the creation of innovative wireless networking systems based on programmable radios. The objectives of this NSF research program are to:

- Capitalize on advances in processing capabilities and radio technology and on new developments in spectrum policy.
- Improve connectivity and make more effective use of shared spectrum resources.
- Enhance the wireless networks community by intermixing the networking, radio, and policy communities, integrating education with research through focused activities, and diversifying participation.

Programmable radio systems offer the opportunity to use dynamic spectrum management techniques to help lower interference, adapt to time-varying local situations, provide greater quality of service, deploy networks and create services rapidly, enhance interoperability, and in general enable innovative and open network architectures through flexible and dynamic connectivity. The specific research areas under this NSF Program are:

- Dynamic spectrum management architectures (including sensor-based architectures) and technologies; includes investigation of issues such as techniques to implement policies, security and robustness, quality of service, and enforcement.
- Topology discovery, optimization, and network self-configuration technology.
- Techniques for the interaction between routing, topology, and administrative/network management including the development of the policy and security framework.

Software Defined Radio Forum

The Software Defined Radio Forum has an embryonic program for investigating technical [9, 10], regulatory, and market aspects of cognitive radio and spectrum efficiency. The technical aspects include the development of security requirements. Some spectral occupancy measurements have been made under the auspices of the SDR Forum. The results of these measurements will be useful in determining the parameters to be used in designing policy-based cognitive radio systems and identifying which portions of the band should be considered for the initial application of such radio systems.

Few ITU-R questions have been related on cognitive radio systems in land mobile and radio termination services. However, no specific Radio Regulation has yet defined cognitive radio. However, in the ITU-R Draft Report, the cognitive radio has been defined as "A radio or system that senses, and is aware of its operational environment and can be trained to dynamically and autonomously adjust its operating parameters accordingly and to learn from the results of its actions and environmental usages pattern". Working Part 8A (Question ITU-R 241/8) has addressed the issue of cognitive radio in land mobile systems and recognized that cognitive radio systems not only have to make more efficient use of spectrum, but also offer more versatility and flexibility, with the increased ability to dap their operations based on external factors. As per the ITU-R Recommendations, the study on the Question ITU-R 241/8is slated for its completion by 2010. It is hoped that by using underutilized spectrum, cognitive radio will provide a 10 times spectrum capacity improvement. At the World Radio Conference (WRC-03), the regulatory community agreed on a method for 5 GHz spectrum sharing of radar and wireless access systems. The basis for the sharing was agreement on the use of Dynamic Frequency Selection in 5230–5350 MHz and 5470–5725 MHz. The WRC-07 shall be addressing cognitive radio issues more effectively and may consider sharing of some other frequency bands.

2.3.7 Spectrum Monitoring for Cognitive Radios

The growing numbers of wireless standards are reducing the amount of unlicensed frequencies, making more and more difficult the use of spectrum for incoming and new wireless communication modes. However, large parts of licensed bands are unused both for a large part of time and space: this means that, even if a particular range of frequencies is reserved for a standard at a particular time and at a particular location it could be free and available. The Federal Communication Commission (FCC) estimates that the variation of use of licensed spectrum ranges from 15 to 85% whereas Defense Advanced Research Projects Agency (DARPA) claims that only 2% of the spectrum is in use in the US at any given moment, even if all bands are allocated. The spectrum measurements taken in Germany and the US given in Figures 2.2a and b indicate specifically the usage of frequency bands in time of the day and the amount of usage for large band of spectrum from some MHz to 6 GHz.

It is then clear that the solution to these problems can be found dynamically looking at spectrum, as a function of time and space.

This is the base of Cognitive Radios (CR): the paradigm, defined first time by Mitola [8], foresees devices able to adapt themselves to spectrum environment and, in general, to external environment and to learn, as a biological cognitive process, from experience how to carry out this adaptation. "CR adjusts its own footprint in accordance with channel foot print". Spectrum sensing enables CR to sense the band occupancy by legacy users. CR users can use the band in absence of legacy user with minimum specified amount of interference to the legacy users. Reconfigurability of CR allows for changing the radio parameters such as transmission power, operating frequency, modulation scheme and the communication technology on the fly without changing any hardware components. CR brings to the definition of a completely adaptable physical layer where communication features, by sensing the spectrum, can change in relation to the conditions of the wireless channel, to the traffic status and to the users' requirements. In this process, in order to allow a representation of the external environment as close as possible to real world, a key role is played by mode identification and spectrum monitoring (MISM). By using MISM the terminal collects fundamental data from external environment, in particular from radio channel, and can carry out the adaptation typical of CR.

MISM is the join of a qualitative and quantitative analysis of reference band through the collection of information in terms of, respectively:

- frequencies usage;
- air interfaces classification.

To evaluate the use of frequencies in a particular band some parameters have been studied, and energy level and interference temperature [3] are the most used; both qualitatively describe with good performance the occupation of a given frequency band. Whereas to provide a quantitative description of spectrum, air interfaces classification (also called mode identification) is performed: an air interface (also called transmission mode) can be defined as the specification of the radio transmission between a transmitter and a receiver. It defines the frequencies or the bandwidth of the radio channels, and the encoding methods used such as FH-CDMA, DS-CDMA, TDMA, MC-CDMA, etc. [4]; thus mode identification process says which standard is present providing data about its nature.

Cognitive radio technology can also simplify regulatory policy. Since users can share spectrum, the demand for special use licenses may be reduced.

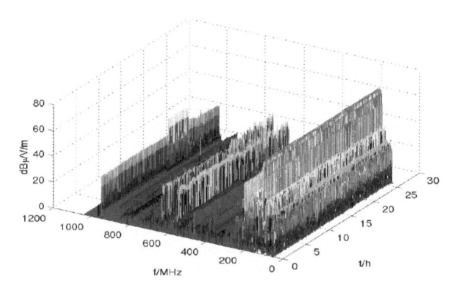

(a) Daily electromagnetic field strength observations, Lichtenau, Germany [1]

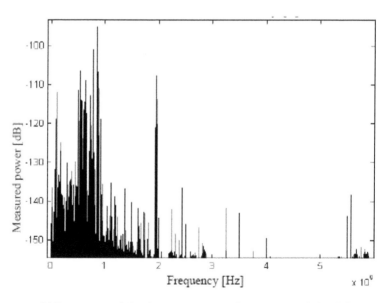

(b) Power spectral density measurements, Downtown Berkeley, US

Figure 2.2 Measurements of spectrum utilization (adapted from [12])

The cognitive radio technologies could be required to meet ever increasing requirements for spectral efficiency, effective etiquette, and resistance to interference. It is noteworthy that the cognitive radio network concept has the potential to explicitly protect the spectrum rights of incumbent license holders. A good example of this is the potential for shared use of spectrum reserved today for military use. In most parts of the geographical locations, this spectrum goes unused most of the time, making it ideal for shared use spectrum. However, in the event of an emergency, a cognitive radio network based system could easily evacuate the full band on a nationwide level for immediate and exclusive military use. This ability is also critical to cellular operators, where the large additional spectrum pool, dramatically reduced costs, and ability to guarantee service through use of incumbent licensed spectrum makes the cognitive radio based network very attractive. Cognitive radio offers hope to meet this demand with a system that is compatible with existing deployed wireless systems, stimulates new innovation, reduces regulatory burden, encourages market competition, preserves the rights of incumbent spectrum license holders, and benefits the populace overall. Therefore, for the purpose of effectiveness the frequency bands are required to be continuously monitored.

Process of Spectrum Monitoring

Three key concepts of spectrum monitoring are:

- Validating information on legitimate users.
- Evaluating real levels of usage of the spectrum.
- Identifying areas for further use, sharing or reallocation.

Classes of spectrum monitoring:

- Monitoring band occupancy:
  At the allocation stage, when existing band undergoes a change of use or when a new service is introduced. To ensure that the old services have vacated the band or that any sharing arrangements are in fact working in practice.
- Monitoring channel occupancy:
  This will confirm or detect the level of use of frequency assignments in particular channels.
- Interference investigation:
  This deals with the discovering and resolving and complaints regarding interference of international interference sources.

What is being monitored?

- Fixed monitoring and HF direction finding.
- VHF/UHF mobile and remote monitoring.
- AM, FM, television, and microwave monitoring.
- Satellite monitoring.

Spectrum monitoring issues are:

○ Regulatory issues: Monitoring the compliance with the conditions in the license.
○ Interference issues: Resolving harmful interference cases.

2.4 Key Research Challenges

- Find average spectrum occupancy in different bands starting from HF to 100 GHz (including protected military bands.
- Feasibility study on selection of competing technologies for CR trials in context to backward compatibility.
- Decide which bands would be available for initial CR trials.
- Develop cognitive radio algorithms to enable co-existence between IEEE 802.11b and 802.16a networks in the same unlicensed band.
- Develop two wireless systems which share radio resources in frequency, space and time, and coordination methods to be used to reduce the mutual interference.
- The effect of interference produced by the man-made/natural environment.
- Define "harmful interference" within the authorized CR bands.
- Investigate how radiated RF power at its location varies with distance along the ground, among obstructions and up in the air.
- Develop practical algorithms for the spectrum-sensing for cognitive radio, for exchanging spectrum-monitoring data between cooperating communications nodes.
- Mapping of the local RF environment of low, medium and high bands, mapping as a function of space, time and frequency of propagation.
- Study the complex multipath components – thereby accounting for reflections of signals from objects such as buildings and vehicles.
- Define the rights of the licensed spectrum band users and on interference protection given to the legacy users where CR technology is authorized.

2.5 Question ITU-R 241/8

Cognitive radio systems in the mobile service

Considering

- That the use of mobile radio systems is growing at a rapid rate globally.
- That more efficient use of spectrum is essential to the continued growth of such systems.
- Those cognitive radio systems may facilitate the more efficient use of spectrum in mobile radio systems.
- Those cognitive radio systems may offer functional and operational versatility and flexibility in mobile radio systems.
- That considerable research and development is being carried out on cognitive radio systems and related radio technologies.
- That the implementation of cognitive radio systems may include technical and regulatory issues and it is beneficial to identify the technical and operational characteristics.
- That Reports and/or Recommendations on cognitive radio systems would be complementary to other ITUR Recommendations on mobile radio systems.

Noting

That there are network aspects related to the control of cognitive radio systems,

It should be decided that the following questions should be studied:

- What is the ITU definition of cognitive radio systems?
- What are the closely related radio technologies (e.g. smart radio, reconfigurable radio, policy-defined adaptive radio and their associated control mechanisms) and their functionalities that may be a part of cognitive radio systems?
- What key technical characteristics, requirements, performance and benefits are associated with the implementation of cognitive radio systems?
- What are the potential applications of cognitive radio systems and their impact on spectrum management?
- What are the operational implications (including privacy and authentication) of cognitive radio systems?

- What are the cognitive capabilities that could facilitate coexistence with existing systems in the mobile service and in other radio communication services, such as broadcast, mobile satellite or fixed?
- What spectrum-sharing techniques can be used to implement cognitive radio systems to ensure coexistence with other users?
- How can cognitive radio systems promote the efficient use of radio resources?

Further decisions to be taken:

- That the results of the above studies should be included in one or more recommendations and/or reports;
- That the above studies should be completed by the year 2010.

2.6 Spectrum Issues for Mobile Broadband Services Like LTE and WiMAX

Operators around the world are demanding mobile broadband services. So much so that they are now seeking an aggressive capacity-easing move to Long Term Evolution (LTE) technology, all-IP data-optimized technology that offers ultra-fast broadband services. While the business proposition surrounding the move to LTE is clear, the spectrum operators' need to make that transition is less so. Radio spectrum is considered the lifeblood of every mobile operator, and today, this finite resource is quickly reaching its limit because of the heavy demand for high-speed data services.

Regulators in many regions of the world are realizing the need to release new spectrum to unleash broadband services that are swiftly becoming so pervasive. "Mobile data is growing fast. In some places, operators experienced a 14-fold increase during 2007. Many operators are reporting 3.5G networks are running at close to capacity during peak hours despite having deployed their available UMTS carriers," noted Darren McQueen, vice president wireless broadband product management with Motorola. "Competitive flat rate pricing and easy-to-use Web-friendly smart phones are turning many customers into heavy users of service. These trends are causing operators to take a closer look at deploying LTE technology as soon as it is available in order to relieve the capacity crunch they may face with today's 3.5G networks. The gating factor, however, is how fast regulators can release new spectrum in time for LTE commercial network availability at the end of 2009."

In the US, for example, regulators have already released commercial spectrum suitable for LTE services, which means operators in this region will become one of the first LTE operators. The Federal Communications Commission (FCC) in April 2008 auctioned 62 megahertz of spectrum in the 700 MHz band. The band is highly prized because the low frequency allows signals to travel farther and provide better in-building coverage than higher frequencies such as 2.1 GHz. As a result, operators need fewer base stations to cover an area, which translates into lower overhead costs – a major asset for any operator looking to be aggressive on the pricing front.

Following the US lead, indications are strong that Europe and many other regulators will release a significant amount of spectrum in the UHF band as broadcasters in various regions also vacate analog TV airwaves and transit to digital TV. Spectrum in the 2.6 GHz band is also coming up for grabs throughout Europe and as such is likely to be the first band for LTE deployments. As much as 140 megahertz of the spectrum (2×70 MHz) will be allocated for FDD services like LTE. To date, Norway and Sweden have auctioned spectrum while the Netherlands, Germany, Austria and the UK have auctions planned. The 2.6 GHz band combined with the digital dividend 700 MHz band will provide an excellent complement for operators. The 2.6 GHz spectrum offers the potential for greater capacity and lower frequency spectrum enables better propagation.

The core UMTS band at 2.1 GHz will also see LTE deployments, especially in Asia, where Japan's NTT DoCoMo and KDDI will roll out LTE services. Some 150 countries are using the band for WCDMA services, but many have not utilized the entire spectrum in the 2.1 GHz band, paving the way for potential LTE deployments alongside WCDMA networks. In all, given LTE's spectrum flexibility, the technology has a bright future since it literally can be squeezed into a number of spectrum bands. Simultaneously, the WiMAX systems are also becoming available in 2.5 GHz as well as 700 MHz bands for deployment. The WiMAX systems also offer spectrum efficiency and data speeds similar to promised by LTE.

- LTE (Long Term Evolution) is the last step toward the 4th generation of radio technologies designed to increase the capacity and speed of mobile telephone networks.
- Where the current generation of mobile telecommunication networks are collectively known as 3G, LTE is 3.9G.
- LTE is a set of enhancements to the Universal Mobile Telecommunications System (UMTS) which developed by the Third Generation Partnership Project (3GPP), an industry trade group.

- The goals for LTE include improving spectral efficiency, lowering costs, improving services, making use of new spectrum and reformed spectrum opportunities, and better integration with other open standards. LTE is the likely wireless broadband technologies of choice for most users.
- LTE could allow data transfer rates to and from mobile devices between 15 and 100 times faster than 3G networks.
- LTE provides downlink peak rates of at least 100 Mbit/s, 50 Mbps in the uplink and RAN (Radio Access Network) round-trip times of less than 10 ms.
- LTE supports flexible carrier bandwidths, from 1.4 MHz up to 20 MHz as well as both FDD (Frequency Division Duplexing) and TDD (Time Division Duplex), meaning LTE, more than any other technology, already meets key 4G requirements.
- LTE incorporates Multiple in Multiple out (MIMO) technology in combination with Orthogonal Frequency Division Multiple Access (OFDMA) in the downlink and Single Carrier FDMA in the uplink to provide high levels of spectral efficiency and enables cost-efficient solutions for wide carriers with high peak rates.
- Although both LTE and WiMAX use the OFDMA air interface, LTE has the advantage of being backwards compatible with existing GSM and HSPA networks, enabling mobile operators deploying LTE to continue to provide a seamless service across LTE and existing deployed networks.
- The upper layers of LTE are based upon TCP/IP, which will likely result in an all-IP network similar to the current state of wired communications. LTE will support mixed data, voice, video and messaging traffic.
- The main advantages with LTE are high throughput, low latency, plug and play, FDD and TDD in the same platform, improved end-user experience and simple architecture resulting in low operating expenditures. LTE will also support seamless passing to cell towers with older network technology such as GSM, CDMA-One, and CDMA2000 & W-CDMA (UMTS). It also supports both paired and unpaired spectrum. LTE is well positioned to meet the requirements of next-generation mobile networks, both for existing 3GPP & 3GPP2 operators. It will enable operators to offer high-performance, mass market mobile broadband services, through a combination of high bit-rates and system throughput, in both the uplink and downlink and with low latency.

- LTE is in the early stages of its evolution, but several of the world's largest mobile operators have announced trials beginning in 2009 and initial deployments as early as 2010 and 2011. The future for LTE and its already impressive ecosystem is evidence of a well-defined standard and there are already more than 100 operators worldwide that have plans for LTE in their future. Some of the frequency bands proposed to be used for LTE and/or WiMAX systems is given below, along with the countries/regions which have indicated their possible deployment in near future.

References

[1] M. McHenry, NSF spectrum occupancy measurements project summary. August 15, 2005. Available online: http://www.sharedspectrum.com/measurements/#

[2] T.W. Hazlett and C. Bazelon, Market allocation of radio spectrum. In *Proceedings of ITU Workshop Market Mechanism for Spectrum Management*, Geneva, 22–23 January 2007.

[3] Annual Report 2008-09, Telecom Regulatory Authority of India.

[4] ITU survey on radio spectrum management. In *Proceeding sof ITU Workshop on Market Mechanism for Spectrum Management*, Geneva, 22–23 January 2007. http://www.itu.int/spectrum.

[5] K. Sridhara, A. Chandra, and P.S.M. Tripathi, Spectrum challenges and solutions by cognitive radio: An overview. *Wireless of Personal Communications*, 45(3):281–291, May 2008.

[6] I.F. Akyildiz, W.-Y. Lee, M.C. Vuran, and S. Mohanty, Next generation/dynamic spectrum access/cognitive radio wireless networks: A survey. *Computer Networks*, 50:2127–2159, 2006.

[7] C. Doyale, The pricing of radio spectrum: using incentives mechanisms to achieve efficiency. In *Proceedings of ITU Workshop on Market Mechanism for Spectrum Management*, Geneva, 22–23 January 2007, http://www.itu.int/spectrum.

[8] J. Mitola III, Cognitive radio: An integrated agent architecture for software defined radio. PhD Thesis, KTH Royal Institute of Technology, Stockholm, Sweden, 2000.

[9] R.S. Wu, White Paper, IEEE P802.22-09/113r0.

[10] Software defined radio-terms, trends and perspectives, White Paper, Center for Software Defined radio, Department of Electronic Science, Aalborg University.

[11] Software Defined Forum, www.sdrforum.org.

[12] F.B. Frederiksen, Improving spectral capacity and wireless network coverage by cognitive radio technology and relay nodes in cellular systems. *Wireless Personal Communications* 45(3):355–368, May 2008.

Biographies

K. Sridhara, an Indian Telecom Service officer who graduated in 1971, did his B.E. (Hons) in Electronics & Communication Engineering in 1970 and post-graduation in Engineering in Computer technology in 1972 from College of Engineering from Guindy, Chennai with distinction. Later, he did his M.B.A. (Finance) from FMS, Delhi, L.L.B from Delhi University and CFA. During his career in Indian telecom, spanning over 38 years, Mr. Sridhara has held various positions in DoT covering research & development, standardization, network planning, operation and maintenance, office automation and computerization of telecom network, costing, setting up of production facility, finance advice, policy planning, licensing, spectrum management, induction of new technology and regulatory affairs and his entire career is marked by many milestones. In his responsibility as Member (Technology), Department of Telecom, the apex body of the government of India responsible for framing telecom policy, he is associated with licensing telecom services, spectrum management, research & development, standardization, interface with international bodies, policy regarding induction of new technologies, formulation of network security policy and co-ordination with service providers on security matters, regulatory affairs, etc. Mr. Sridhara has traveled widely. He has represented the government of India at various international forums and has made very effective contributions. He is at present Director General, Global ICT Standardization Forum of India (GISFI). Currently, he is keenly involved in the introduction of new generation technologies in Indian telecom network and in framing WiMax policy.

P.K. Garg is a member of the Radio Regulations Board of the ITU, Geneva, and a former wireless advisor of the government of India, having retired after a long service in the field of national RF spectrum regulation and management. He is also a member of the government committee on spectrum allocation procedure and pricing. He has widely traveled since 1980, for participation in various international and regional conferences, meeting and technical study programs of ITU, Asia Pacific Telecommunity (APT), and other international and regional bodies connected with telecommunications. He has also represented India on the Governing Council of ITU, Geneva. He has also served as the Senior UN & ITU Expert in Radio Frequency Spectrum Management in Saudi Arabia. He has held a degree in Electronics and Communication Engineering from Punjab Engineering College, Chandigarh.

He is a Fellow of the Institution of Electronics & Telecommunication Engineers (India) for the last 20 years and is a member of its Governing Council also. He has also been awarded the "Eminent Engineer Award" for the year 2003 by the Institution of Engineers, Delhi Center.

Ashok Chandra, an Indian Engineering Services officer (1976), did his post graduate in Electronics and PhD in Electronics and Doctorate of Science (D.Sc.) in Radio Mobile Communications. He has also worked as a guest scientist at the Institute of High Frequency technology, technical University (RWTH), Aachen, Germany during 1995 and 1999. In 2002, he worked as a guest scientist at Bremen University, Bremen, Germany, where he undertook a series of research studies in the area of radio mobile communications. Dr. Ashok Chandra has more than 30 years of technical experience in the field of radio communications/radio spectrum management including about seven years dealing with Technical Education matters of the Indian Institutes of Technology, Indian Institute of Science, etc., particularly their various research projects, also in the areas of telematics/radio communications, and delivered over 25 research presentations at International Conferences in the areas of EMI, Radio Propagation, etc. He has visited various technical institutions and universities, such as Technical University of Aachen, Germany, Aalborg University, Denmark, Bremen University, Germany, and University of Lisbon, Portugal, and took several lectures in the area of radio mobile communication at Bremen University and Aalborg University. He has chaired various technical sessions at the International Conferences on Wireless Personal Communications and he is on the panel of reviewers of the international journal *Wireless Personal Communications*. In his responsibility as the wireless advisor to the government of India, he is associated with spectrum management activities, including spectrum planning and engineering, frequency assignment, frequency coordination, spectrum monitoring, policy regarding regulatory affairs for new technologies and related research & development activities, etc. He is also associated with the implementation of various prestigious World Bank Projects on National Radio Spectrum Management and Monitoring System (NRSMMS). These projects include automation of Spectrum Management processes and design, supply, installation/commissioning of HF/VHF/UHF fixed monitoring stations; V/UHF mobile monitoring stations, and LAN/WAN communications network, etc. He was elected Vice-Chairman, Study Group 5 of ITU-Radio Sector. His areas of interest include radio regulatory affairs for new technologies, radio mobile communications and cognitive radio.

P.S.M. Tripathi, an Indian Engineering Services officer who graduated in 1985, did his B.E. in Electronics from M.M.M Engineering College, Gorakhpur, India. He has over 10 years of technical experience in the filed of Radio Communications/Radio Spectrum Management. In his current responsibility as engineer in the Wireless Planning & Coordination Wing of the Department of Telecommunications, at the Ministry of Communications & IT of the government of India, he is associated with spectrum management activities, including spectrum planning and engineering and policy regarding regulatory affairs for new technologies and related research & development activities and ITU-R related matters. He has also been associated with implementation of a very prestigious World Bank Project on National Radio Spectrum Management and Monitoring System (NRSMMS). This project includes automation of Spectrum Management processes and design, supply, installation/commissioning of HF/VHF/UHF fixed monitoring stations; V/UHF mobile monitoring stations, and LAN/WAN communications networks, etc. His area of interest includes radio regulatory affairs for new technologies and cognitive radio. Presently he is working as a research fellow in department of electronics, University of Tor Vergata, Rome, Italy.

3

Future Radio Access Technologies

B.A. Damahe[1], Olive Jesudas[1] and Girish Mangalani[2]

[1]*L&T, Mumbai, India*
[2]*Reliance Communications, Mumbai, India*

This chapter presents the discussion on the Future Radio Access Technologies. Initially the Requirements and Challenges for Future Radio Access Technologies are discussed with respect to Data Rate, Spectrum, Energy Efficiency and Advanced Channel Models. The need for improving the Spectrum Efficiency is discussed ahead with Multiple Antenna Systems, Relaying and Base Station Cooperation, Heterogeneous Network Environments. At the end, Software Defined Radio (SDR) and Cognitive Radio (CR) are discussed.

The chapter is organized as follows. Section 3.1 covers the challenges for future radio access technologies including data rate requirements as well spectrum utilization and energy efficiency issues. Section 3.2 outlines advanced schedulers and measurement schemes and the need to improve spectrum efficiency with respect to heterogeneous networks. Section 3.3 describes Emerging Trends such as MIMO and Cooperative relaying in CWC. Section 3.4 elaborates the vehicular networks and multi radio access challenges. Section 3.5 deals with the architecture of software defined radio, its software structure, applications and advancements. Section 3.6 depicts the architecture, design and main functionalities of SDR as well as spectrum sensing applications and benefits of CR. Section 3.7 explains the evolutions of ITU-T standards. Section 3.8 covers the Femto cell technology with its advantages, standardization and 3GPP specifications for it. Section 3.9 concludes the chapter while some future challenges are mentioned in Section 3.10.

R. Prasad (ed.), Future Trends and Challenges for ICT Standardization, 75–105.
© 2010 *River Publishers. All rights reserved.*

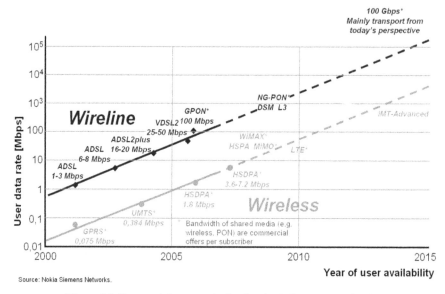

Figure 3.1 Expected data rates in fixed and wireless communication

3.1 Challenges

A key requirement on a new Radio Access Technology is the provision of significantly enhanced services due to IoT with improved QoS by means of higher data rates, lower latency, spectrum reuse, power efficiency, etc., at a substantially reduced cost, compared to current mobile-communication systems and their direct evolution.

3.1.1 Data Rate Requirements

The expected data rates in fixed and wireless communication systems are shown in Figure 3.1. The exponential growth of the data rates in wireless systems has followed the one experienced in fixed (wire line) systems with a time offset of less than five years. The next generation networks will support high data rate applications like the next generation (3D) Internet with 3D graphics and animations.

The Wireless World Research Forum expects that up to 7 trillion wireless devices will serve up to 7 billion people by the year 2017. Thereby, people will always be connected to the Internet. As a consequence, we will run out of usable spectrum, and an increased spectral efficiency as well as more intelligent ways of accessing and allocating the spectrum is required.

3.1.2 Spectrum Requirements and Challenges

Presently in India, the following 3G spectrums are available.

1. 2.1 GHz:
   - 1920–1980 MHz,
   - 2110–2170 MHz.
2. 2.5–2.69 GHz:
   - 2500–2570 MHz,
   - 2620–2690 MHz.

Recently, the World Radio Communications Conference 2007 (WRC-07) has identified the following frequency bands for IMT applications that include Third Generation (3G) and future Fourth Generation (4G) wireless communication systems:

- 450–470 MHz,
- 698–806 MHz,
- 2.3–2.4 GHz,
- 3.4–3.6 GHz.

Due to these decisions of the WRC-07 on the identification of new spectrum for IMT, it will be very difficult to obtain dedicated IMT-Advanced spectrum for several operators in a given area based on the identified bands, since the anticipated high data rates of IMT-Advanced require a large bandwidth per operator. The additional radio spectrum assigned for mobile communications is in relatively small fragmented bands with some remaining uncertainty concerning the availability of these bands. Therefore, new adaptive spectrum sharing models have to be designed and the spectral efficiency of future wireless radio systems has to be increased.

The identified consecutive bands divided by the necessary carrier bandwidth per operator may not result in a sufficient number of operators to support competition. Therefore, infrastructure sharing may be required to support a quality of service with sufficiently wide carrier bandwidths and competition between different operators.

The natural resources frequency spectrum and energy are limited. The higher the carrier frequency becomes, the more difficult are the propagation conditions. Doubling, for example, the carrier frequency in an environment that has a path loss exponent (R^α) $\alpha = 3$, results in a 9 dB higher propagation loss (1/8 of the power), the cell radius shrinks to 50%, the cell area shrinks by 25%, and 300% more base stations are required to cover a given area, which results in a three times higher investment for operators as well as more energy consumption. Therefore, spectrum in the low frequency range would be beneficial for cellular communications. Due to higher frequencies more efficient topologies and much more efficient transmission techniques are needed.

Lower Frequency Range Spectrum

Spectrum at 450 MHz, is associated with lower propagation loss and, therefore, a better coverage. Typically, there are smaller spectrum slices, thus the capacity is smaller. Lower Doppler frequencies provide support for a higher mobility, e.g., fast trains. Moreover, the lower transmit power leads to lower emissions and lower power consumption. Thus, the spectrum in the lower frequency range is most valuable from a coverage point of view.

Higher Frequency Range Spectrum

Spectrum in the higher frequency range, e.g., at 3.5 GHz, is associated with higher propagation losses. Therefore, the coverage is worse. Typically there are larger spectrum slices. Therefore, the available capacity is larger. To provide coverage, a higher transmits power or smaller cells are required, which leads to larger emissions and a higher power consumption. Thus, the spectrum in the higher frequency range is most valuable from capacity point of view (since a larger part of the spectrum is available in this range).

3.1.3 Coverage Limited Scenarios

Either in an indoor environment or with big reuse clusters providers can cut the cost and increase the service data rates by dynamic spectrum management to enable multiple operators (e.g., with different profiles like business and leisure) in the scarce frequency bands at low carrier frequencies. This enables high peak user data rates and a competition among service providers also at low carrier frequencies. New business models have to be developed (with supporting infrastructure) were the unused spectrum can be traded. Moreover, this new business model has to give incentives to be spectrally efficient also at low loads for a particular operator.

3.1.4 Capacity Limited Scenarios

In the interference-limited case, operators might use higher carrier frequencies in the multi-GHz range, e.g., at 20–30 GHz. Here, potentially a wider spectrum is available. In this case, however, only a small range can be covered, where unobstructed line-of-sight might be required. Also in this scenario, dynamic spectrum management leads to new business models and therefore needs supporting infrastructure for spectrum trading. A large bandwidth per operator enables high spectrum efficiency due to multi-user scheduling, link adaptation, and adaptive antenna gains. Reliable services can be realized due to fast retransmission schemes. Moreover, the unused spectrum can generate revenues from spectrum trading. Future local area operators might include traditional telecommunications operators, local operators (e.g., shopping malls, office buildings), and micro operators.

3.1.5 Multi-GHz Carrier Frequencies

Challenges to be overcome at multi-GHz carrier frequencies include the Radio Frequency (RF) front- end (co-design the physical layer design with the RF hardware to mitigate impairments). In short range scenarios, the high carrier frequencies enable a dense spatial frequency reuse with many nodes (higher capacity per area). If multiple hops are used, only a few base station sites are required. In this case, quality of service (QoS) requirements on short delays can still be met with a wide system bandwidth. This, however, requires a fast digital hardware, where parallel physical layer algorithms are implemented. Furthermore, the corresponding channel models at high frequencies with rich spatial information have to be developed.

3.1.6 Multiple Hops at Higher Carrier Frequencies

Challenges and opportunities with multiple hops at high carrier frequencies include new cellular structures with cooperation to realize macro-diversity gains. In the field of wireless communication, Macro-diversity is a kind of diversity gain where several receiver antennas and/or transmitter antennas are used for transferring the same signal. In this scenario, the capacity will be more smooth as a function of the cell radius and movable to where users are instantaneously operating (moving cell structures and networks). Moreover, the system should be able to coordinate and collaborate over multiple "cells" to achieve macro-diversity gains. It should also adapt in three dimensions in dense metropolitan areas using advanced channel knowledge and integrate peer-to-peer cooperation. The multiple nodes should be realized via very cheap hardware. There will be hybrid networks, where the core infrastructure (that guarantees a minimum quality of service) is owned by the operator, but with user incentives and network support for user added nodes. Resource management schemes should be developed to control the quality of service along with supporting business models. Reliable and efficient handover schemes can be realized by predicting the user mobility. Moreover, there are opportunities for moving networks with group handover as well as multicast support.

3.1.7 Energy Efficiency

The transmitted data volume increases approximately by a factor of 10 every 5 years, which corresponds to an increase of the associated energy consumption by approximately 16–20% per year. Currently, 3% of the world-wide energy is consumed by the ICT infrastructure which causes about 2% of the world-wide CO2 emissions (which is comparable to the world-wide CO_2 emissions by airplanes or one quarter of the world-wide CO_2 emissions by cars). If this energy consumption is doubled every 5 years, serious problems will arise. Therefore, the energy consumption of future wireless radio systems should be reduced, which creates inter-disciplinary research challenges including semiconductor technology, hardware, networks, services, and radio transmission, where schemes have to be designed that operate with a reduced transmit power.

Energy Efficient Mobile Computing

Energy efficient mobile computing requires energy efficiency in the user terminals for mobile computing at high carrier frequencies to save battery life time. Short hops are important for low peak power requirement. Nevertheless, antenna array and macro-diversity gains can be realized. A co-design of the hardware and the physical layer is required to mitigate hardware impairments.

3.2 Advanced Schedulers and Measurement Schemes

Advanced schedulers and measurement schemes should predict the channels, based on rich channel model information, to transmit with high instantaneous data rates and mainly on good resources. Moreover, the buffer levels should also be predicted. Thus the terminals can sleep as much as possible to save energy.

3.2.1 Long Term Scheduling Schemes

Long term scheduling schemes are important to decide when to transmit and when to receive. For non-real-time services, service latency is the most important parameter for user experience. As memory is constantly becoming more cost efficient (with a smaller footprint), the user terminal should act as a cache for pre-fetched data based on predicted user service consumption patterns. Terminals can receive pre-fetched data during off-peak hours and instantly download increments only. This enables a time-aligned transmission of popular services to multiple users on the downlink using multicast and might be facilitated by peer-to-peer transmission of cached data.

3.2.2 Advanced Channel Models

To be able to design and evaluate such energy efficient, adaptive wireless systems that are more sensitive to the environment and use sophisticated scheduling schemes, advanced models are required to model the channel behavior realistically in time, frequency, and space. Therefore, the dynamics of the channel have to be investigated and modeled. Depending on the environment and the carrier frequency, stationary regions have to be found in space and time. Also channel models for new types of networks, such as wireless sensor networks, body area networks, and car-to-car communications, have to be developed. Moreover, adequate models for the prevailing channel state information and multi-user interference in multiple antenna systems are needed, e.g., MIMO systems that interfere with one another. Here, also the role of diffuse multi-path has to be investigated. The evaluation of cognitive radio systems also requires channel models of the same environment on different carrier frequencies at the same time and at the same location.

3.2.3 Need for Improving the Spectrum Efficiency

The higher desired data rate calls for higher spectral efficiency since the spectrum is a finite and fixed resource. The spectral efficiency [5] is traditionally given by bits per Hz of channel bandwidth. But for realization of Indian Mobile Telecommunications-Advanced (IMT-Advanced) cognitive radio is the answer which has the ability to reuse spectrum opportunistically with minimal interference to the legacy users. This will enhance the spectral efficiency in real terms. Shannon says that spectral efficiency can be made infinite but at the cost of exponential rise in energy per bit of efficiency. In the CR paradigm, interference to the legacy users by secondary (unlicensed) users is very important factor which calls for another definition of spectrum efficiency. In general, raising the spectral efficiency will raise the energy radiated per bit thereby reducing the number of bits per unit spatial area. Hence spectral efficiency in terms of bit per unit area is important in CR context.

As compared to current wireless radio systems, the spectral efficiency of future systems has to be improved significantly to reduce the cost of these wireless radio systems. This can be achieved via advanced signal processing algorithms (interference cancellation, efficient signal processing for multiple antenna systems, etc.). It took 20 years of engineering research (1987–2007) to improve the spectral efficiency of GSM (0.1 bits/s/Hz/sector) by a factor of approximately ten to get LTE with a spectral efficiency of about 1.2 bits/s/Hz/sector. Therefore, more research is required to get further improvements that might be achieved in the areas of multiple antenna systems, relaying and base station cooperation, heterogeneous network environments as well as cognitive radios. Note that cognitive radios as such do not improve the spectral efficiency. Such systems may use additional frequency bands, which would otherwise not be used. (In fact, more spectrum than the originally identified bands would be used.) A key concept would be to mitigate the impact of intra- and inter-cell interference, because in the noise-limited case modulation and coding schemes are already rather close at the Shannon bound.

3.3 Emerging Trends

The emerging trends such as MIMO smart antennas, Cooperative Wireless Communication, Vehicular Communication,, Heterogeneous networks as well as Femto cells are playing the vital role in the Telecom field.

3.3.1 Multiple Antenna Systems

The adoption of multiple antenna techniques in future wireless systems is expected to have a significant impact on the efficient use of the spectrum, on the minimization of the cost of establishing new wireless networks, on the optimization of the service quality provided by wireless networks, and the realization of transparent operation across multi-technology wireless networks. Nevertheless, the design of future generation smart antenna systems involves a number of challenges, such as reconfigurability to varying scenarios in terms of propagation conditions, traffic models, mobility, transceiver architectures, mobile terminal resources, i.e., battery life time, quality of service requirements for different services, and interference conditions. The design of robust solutions matched to the reliability of the available channel information is required in order to account for the impact of channel estimation errors and feedback quantization and delay. Moreover, the system architecture as well as implementation and complexity limitations need to be taken into account in the design of multiple antenna techniques.

Advanced multiple antenna schemes are a key component to increase the spectral efficiency of wireless communication systems, to provide an efficient coverage of high data rates (also to users on the cell edge) along with an efficient support of extremely high data rates, and to enhance the interference management. Using multiple antennas at the transmitter, we can transmit multiple data streams to individual users (spatial multiplexing) as well as multiple users at the same time, which is referred to as a multi-user MIMO system. Nevertheless, Space Division Multiple Access (SDMA) concepts did not take off, when they were introduced for the first time. This is due to the fact that proper scheduling concepts through cross-layer designs were not available at that time. Moreover, the only available feedback was for power control and did not take into account that channel state information is required at the transmitter (CSIT). In FDD systems with a dedicated feedback channel, the CSI-T quality depends on the feedback rate.

- In FDD system without feedback, a partial CSI-T knowledge may be obtained by estimating reciprocal channel components (angles, delays). Nevertheless, the robustness of such schemes has to be investigated.
- In TDD systems, the beam former may use uplink channel impulse response estimates (since the average channel behavior is reciprocal), where the quality depends on the "ping-pong time", i.e., the time interval between the channel estimation and the downlink transmission (frame structure), and the channel coherence time.

With cooperative communication, the transmission diversity is achieved by enabling a single antenna device from multi-user scenario to share their antennas and generate a virtual multiple antenna transmitter. The transmit diversity methods such as Almouti Signaling have been incorporated into wireless standards is the great acknowledgement for MIMO systems [1].

Open problems in this area include:

- Multi-user MIMO communications in frequency selective channels and in the interference channel.
- Optimum design of *reduced feedback* schemes for multi-user MIMO systems.

- Low complexity downlink transmission strategies.
- The impact of multiple access (scheduling) protocols on the design of multi-user MIMO algorithms as proper scheduling allows for simplified transmitters and simplified MIMO/multiuser detectors those are able to save battery life time.

3.3.2 Relaying and Base Station Cooperation

The cooperation between base stations (multi-cell approaches) and cooperation between mobile terminals can enhance the spectral efficiency significantly. The latter includes relaying between inner cell and outer-cell users. It allows to overcome the attenuation and even to perform distributed beam forming. If multiple layers of the OSI model are treated jointly (cross-layer designs), significant gains can be achieved by integrating, for example, the physical layer and the network layer. First research results already exist for the two-way relay channel. In this case, Joint Network and Channel Coding (JNCC) that combines distributed channel coding and network coding (in conjunction with Turbo Decoding and Hybrid ARQ) can increase the throughput by more than 150% relative to point-to-point transmission using the same energy and bandwidth. The currently emerging field of network coding being still in a theoretical stage should be carefully monitored for its potential to generate applications in cellular mobile systems in order to allow further capacity enhancements. The currently emerging field of network coding, which is still in a theoretical stage, should be monitored carefully for its potential to generate applications in cellular mobile systems in order to allow further capacity enhancements.

The coverage and capacity on the uplink are limited by thermal noise at the base station and intra-cell interference from active users in the same cell and on the same frequency. Moreover, taking into account frequency reuse, the (unpredictable) inter-cell interference from active users in neighboring cells becomes a problem. This can be overcome via base station cooperation with (turbo) interference cancellation using a joint decentralized detector. Simplified schemes have to be designed as a full joint detector would be too complex and the backhaul load on the optical terrestrial link should be limited.

In cooperative communication, the information overheard by neighboring nodes is intelligently used to provide the healthy communication between a source and the destination called a sink. In CWC, several nodes work together to form a virtual array. The overheard information by each neighboring node or relay is transmitted towards the sink concurrently. The cooperation from the wireless sensor nodes that otherwise do not directly contribute in the transmission is intelligently utilized in CWC. The sink node or destination receives numerous editions of the message from the source, and relay(s) and it estimates these inputs to obtain the transmitted data reliably with higher data rates. CWC has the following advantages:

- Higher reliability: lower error probability with more throughput.
- Reduced transmitter power with less interference.
- Provides considerable resistance to small scale as well as shadow fading with high spatial diversity.
- Opportunistic use and reorganization of the network with increased energy efficiency and higher spatial diversity.
- Cooperative transmission assures remarkable improvements in overall sturdiness of WSN and network throughput. It also makes sure a noteworthy reduction in interferences and different delays.
- Coverage range extension is the beauty of CT.
- CWC is resistant to large scale shadowing.

The idea of integration of PHY cooperation with MAC sub layer for the improvements in throughput and interference. For authentication of Coop-MAC with receiver-combining capability design proposal, off-the-shelf IEEE 802.11b network interface cards (NICs) on a Linux platform is utilized [2]. Similar delicate cross-layer design issues proliferated in wireless ad hoc networks, and the connotations of node cooperation, including cooperative routing algorithms and the scalability of network capacity with the number of nodes in a network, deserve further investigation.

3.3.3 Heterogeneous Network Environments

Future wireless networks can be enhanced dramatically by taking serious advantage of the heterogeneous network environment, by applying spectrally more efficient transmission schemes, and by deploying novel network topologies. These include ad hoc networking, mesh networks, peer-to-peer communications, relaying, classical cellular networks, and distributed antennas. A key question is how different types of networks could be managed jointly. The major issues related to different opportunities to increase the intelligence include:

- Opportunistic spectrum use and spectrum sharing.
- Spatial domain utilization to improve the capacity.

- User location information and location awareness.
- User co-operation.
- Existence of multiple radio access technologies.
- Exploitation and acquisition of channel state information and channel statistical behavior information.
- Traffic pattern knowledge including long-term monitoring and prediction as well.
- Network topology awareness in a mobile device.

3.4 Vehicular Networking

Vehicular networking could be thought as one of the first application areas for the new wireless networking concepts to boost also the development of other wireless applications and related technologies. The automation and amount of electronics of cars will continue to increase, and a car is a natural intelligent environment waiting for the full deployment of wireless technologies. In this case, broadband access is provided almost everywhere ("always connected") and accurate position information is available. Context sensitive services will emerge for drivers and passengers on the road to enrich their experience and to improve safety. Moreover, these context sensitive services will also support other parties like car manufacturers. Therefore, driving will become an interactive session with the surrounding real and an available virtual environment. Challenges include the multi-radio access (always-best connectivity), the mobility and the radio channel dynamics, the fast changing sensory information within a vehicle and to the vehicle as well as the data and information exchange and dissemination within a car, between cars, and to the car.

3.5 Software Defined Radio

Software Defined Radio (SDR) is a term which is used for a range of different radio implementations in software. The term SDR is used as the name for the use of software which performs all the signal processing and control functions for the received radio signals as well as the applications processing. The hardware/software partition point in this case is the low-IF or base band I/Q sampled received signals prior to the demodulation stage. Hence all demodulation and quite often the front-end filtering for adjacent-channel rejection purposes is all done in software. Such a structure has many constraints with which it must deal, including the synchronization, detection, and demodulation and decoding functions required for digital communications receivers as well as the system control, user interface and the scheduling of data transfer both within and to/from external sources. On top of this, we require a software system solution which is reliable but which can offer the ability to update software, add new features and progress to new models with a minimum of cost.

An SDR is characterized by its flexibility: Simply modifying or replacing software programs can completely change its functionality. This allows easy upgrade to new modes and improved performance without the need to replace hardware.

An SDR can also be easily modified to accommodate the operating needs of individual applications. There is a distinct difference between a radio that internally uses software for some of its functions and a radio that can be completely redefined in the field through modification of software. The latter is a software-defined radio. SDR devices have the capability to dynamically change transmitter and receiver characteristics without changing hardware. Unlike traditional radio devices, SDR subscriber equipment relies on digital signal processing (DSP) chips for modulation and filtering. An SDR device is envisioned to operate with the existing and evolving land mobile radio (LMR) system by automatically switching operational modes.

SDR technology has the potential to considerably reduce the interoperability gaps that currently exist within the public safety community radio networks. This technology holds the promise of offering advanced features and capabilities not available in current LMR systems. However, SDR technology is still at an early stage of development. There are still many proprietary design, spectrum management, and communication security issues that must be fully addressed.

3.5.1 SDR Architecture

The generic SDR architecture comprises specific functional blocks connected via open interface standards. The SDR architecture supports three specific domains: hand-held, mobile, and base-station (or fixed site). For the purposes of this assessment, only the architectures for hand-held and mobile SDR devices will be identified. The software is implemented by controlling the characteristics of equipment/device subsystems that support scalability and flexible extensions of applications. Modularity is an important factor in the implementation of software applications within open systems.

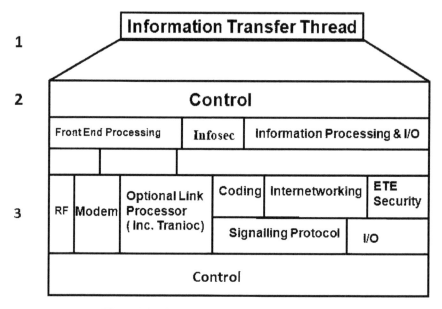

Figure 3.2 Hierarchical functional model of SDR

Figure 3.2 illustrates a high-level hierarchical functional model for a two-way (send and receive) SDR device. Three views of increasing complexity are presented. The top-level view is a simple representation of an entire information transfer thread. The left-hand side interface is the air interface. The right-hand side interface is the user interface.

The next level view identifies a fundamental ordered functional flow of four significant and necessary functional areas:

1. Front end processing.
2. Information security.
3. Information processing.
4. Control.

Front end processing consists of the physical air (or propagation medium) interface, the front-end radio frequency processing, and any frequency up and down conversion. Also, modulation/demodulation processing is contained in this functional block area.

- INFOSEC provides user privacy, authentication, and information protection. In the military and, subsequently, public safety communities, IN-

FOSEC for sensitive and classified communications must be consistent with the government security policies as defined by the NSA.

- Content or information processing is the decomposition or recovery of the embedded information containing data, control, and timing. Content processing and I/O functions map into path selection (including bridging, routing, and gateway), multiplexing, source coding (including coding, and video compression/expansion), signaling protocol, and I/O functions.

The functional components of SDR architecture are connected together via open interfaces. Each functional component in the SDR architecture is controlled with software. The software necessary to operate an SDR device is called a software application. The Software Defined Radio Forum (SDRF) open architecture comprising of seven independent subsystems interconnected by open interfaces. Interfaces exist for linking software application specific modules into each subsystem. Each subsystem contains hardware, firmware, an operating system, and software modules that may be common to more than one application. The application layer is modular, flexible, and software specific. The common software Application Programming Interface (API) layer is typically standardized with common functions based on defined interfaces.

3.5.2 Software Structure

The system is based on a Real-Time Operating Systems (RTOS) which provides a number of services including hard real-time scheduling and memory management. Such RTOSs also usually have a number of associated service applications such as a TCP/IP stack, browser functionality and file system capabilities. In this case, those hard real-time requirements come from the need to receive and process the sampled RF data that is generated by the ADC in the RF system as well as audio and video output signals. Typically, the RF samples enter the processor at a rapid rate, e.g. 8.192 million samples per second for DAB and DAB-derived MTV standards such as DAB-IP and T-DMB.

The software structure shown consists of five primary layers. The software and hardware Kernel layer includes the RTOS and hardware drivers. While the RTOS may be generalized for multiple platforms, it will be customized for each platform due to the specific features of that platform including instruction set, DMA, interrupts and system timers.

Above the kernel layer is the abstraction layer. This is somewhat sandwiched between the kernel layer and the framework layer. When using an RTOS, a number of framework facilities will often be provided by that RTOS. These are usually enhanced by system-specific framework features to ensure that the overall system has the required capabilities. These features may be data-transfer capabilities to assure the real-time data flow between the functional objects, or the provision of timing functions to support synchronized stream applications such as DAB Slide Show. Above the Framework layer is the Applications layer. This in itself may have a number of internal layers. Typically in a multi-standard solution there will be three layers, the demodulation layer which supports the real-time demodulation of the various transmission standards, a transport layer which supports the different transport formats and conditional access methods used and which routes the extracted data to the audio, video and data applications which do the final processing of the received information and present the data to the user through the highest layer, the Human Machine Interface (HMI). The HMI is responsible for both the presentation of the received information but also system control, where it receives commands from the user and then instructs the software system to perform the appropriate instructions to realize those user commands. The software structure described above supports a plug-and-play approach to system implementation. In this case, the system may be minimal for low-cost or even data-pump solutions or it can be fully-featured including a range of audio and/or video functionality. Each customer will have different requirements. The beauty of the SDR approach is its flexibility to provide customized solutions which fulfill the customer's needs in the most cost-effective way.

In this architecture, the framework acts as a connection machine to ensure that data and control information is communicated in an appropriate fashion. All the signal-processing functions are performed in the applications. These applications can be either statically bound at code-link time or dynamically bound during normal system operation. This allows new applications to replace old applications and the reuse of those resources (MIPS, memory, DMA data bandwidth) that were previously occupied. For example, if the user decides to receive DRM instead of DAB, then the DAB stack is unloaded and all the resources it used are released. The DRM stack is then instantiated and it will use part or all of the resources which were previously used for the DAB stack. This flexibility ensures maximum reuse of the system hardware which in turn promotes minimum cost. The plug-and play approach allows the simple addition of new functionality as well as the updating of existing

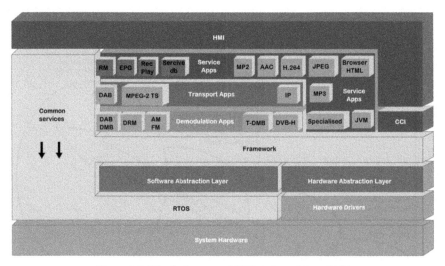

Figure 3.3 Block diagram of SDR (adapted from [11])

functionality. Indeed the structure and hardware abstraction of the code allows it to be deployed on a wide range of hardware platforms. Changes are generally limited to the Abstraction layer and platform-specific optimization using ASM code which is required for speed purposes as the majority of the code is written in C/C++ and is platform-agnostic and hence may be deployed on a range of DSP and RISC (e.g. ARM) platforms.

3.5.3 Applications of SDR

The approach outlined above may be used for basic digital radio systems such as DAB-only or for more highly complex systems such as Mobile TV Sub Systems (MTV-SS) which are multi-standard capable. The block schematic of SDR is shown in Figure 3.3.

3.5.4 Advancement of SDR Technology

Radio technology has progressed from being hardware dependent to being more software oriented. Many people view SDR as the next logical step in the evolution process of radio technology. Figure 3.4 shows the evolution of SDR technology. The evolution of all radio technologies begins with hardware radios. In the future, SDRs will be deployed offering more flexibility and more capabilities. Much farther in the future, the ideal SDR device will

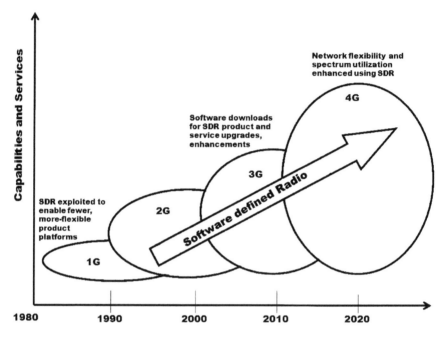

Figure 3.4 SDR technology advancement path (Source: Software Defined Radio Forum)

become available, offering the most flexibility and the most capabilities. The following paragraphs describe the four phases of radio development shown in Figure 3.4.

Hardware Radio

Hardware radios are the baseline for comparison with other radio technologies. Hardware radios were most commonly used in the 1950s and 1960s. These radios were heavy but extremely durable. All radio components were hardwired. Switches, dials, and buttons were the only means for a user to operate the radios. Any changes in operating frequency required physically swapping the crystals that defined the operating frequency of the radio. Typical examples are Bendix King, General Electric, and Motorola radios.

Software Controlled Radio

These radios use modern digital semiconductor technology. The digital integrated circuitry (IC) inside software controlled radios allows for the limited control of functions to be implemented by software. Examples of these control functionalities include frequency reprogramming, changing encryption keys, and programmable switches and buttons. However, software controlled radios cannot change modulation types or operating frequency bands. Most modern LMR radios available today can be classified as software controlled radios.

Software Defined Radio

The majority of the components in an SDR device are operated by software. SDR differs from software controlled radio in that DSP chips are used to generate various modulation types, filters, and air interfaces. However, the RF front end of SDR is still implemented by analog circuitry, which often leads to bulky design, multiple antennas, and architecture complexity.

Ideal Software Defined Radio

Ideal SDRs differs from SDRs in that all radio components are software-driven, including the RF front end. It is expected that ideal SDRs will have a dramatic improvement in overall system performance relative to the initial generations SDRs. However, because of technology limitations, ideal SDRs are unachievable today and may not be realizable in the foreseeable future.

3.6 Cognitive Radio

Today's wireless networks are characterized by a fixed spectrum assignment policy. However, a large portion of the assigned spectrum is used sporadically and geographical variations in the utilization of assigned spectrum ranges from 15 to 85% with a high variance in time [3]. The limited available spectrum and the inefficiency in the spectrum usage necessitate a new communication paradigm to exploit the existing wireless spectrum opportunistically. This new networking paradigm is referred to as cognitive radio networks.

Cognitive radio is a paradigm for wireless communication in which either network or wireless node itself changes particular transmission or reception parameters based on interaction with the environment in which it operates to execute its tasks efficiently. This parameter alteration is based on observations of several factors from external and internal cognitive radio environment, such as radio frequency spectrum, user behavior, and network state. It is also useful to provide highly reliable communication for all users of the network and to facilitate efficient utilization of the radio spectrum in a fair-minded

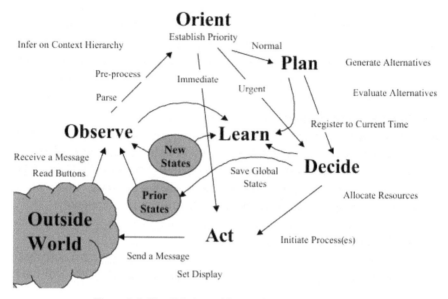

Figure 3.5 Simplified cognition cycle (adapted from [4])

way. A cognition cycle by which a cognitive radio may interact with the environment is illustrated in Figure 3.5.

3.6.1 Cognitive Radio Technology

Although cognitive radio was initially thought of as a software-defined radio extension (Full Cognitive Radio), most of the research work currently focuses on Spectrum Sensing Cognitive Radio, particularly in the TV bands. The essential problem of Spectrum Sensing Cognitive Radio is in designing high quality spectrum sensing devices and algorithms for exchanging spectrum sensing data between nodes. It has been shown that a simple energy detector cannot guarantee the accurate detection of signal presence, calling for more sophisticated spectrum sensing techniques and requiring information about spectrum sensing to be exchanged between nodes regularly. Increasing the number of cooperating sensing nodes decreases the probability of false detection.

Filling free radio frequency bands adaptively using OFDMA is a possible approach. Timo A. Weiss and Friedrich K. Jondral of the University of Karlsruhe proposed a Spectrum Pooling system in which free bands sensed by nodes were immediately filled by OFDMA sub bands.

3.6.2 Applications of Spectrum Sensing Cognitive Radio

Application of spectrum sensing cognitive radio includes emergency networks and WLAN higher throughput and transmission distance extensions. Evolution of Cognitive Radio toward Cognitive Networks is under process, in which Cognitive Wireless Mesh Network (e.g. CogMesh) is considered as one of the enabling candidates aiming at realizing this paradigm change.

3.6.3 Cognitive Radio Architecture and Design

The network-centric cognitive radio architecture under consideration in this project is aimed at providing a high-performance platform for experimentation with various adaptive wireless network protocols ranging from simple etiquettes to more complex ad-hoc collaboration. The basic design provides for fast RF scanning capability, an agile RF transceiver working over a range of frequency bands, a software-defined radio modem capable of supporting a variety of waveforms including OFDM and DSSS/QPSK, a packet processing engine for protocol and routing functionality, and a general purpose processor for implementation of spectrum etiquette policies and algorithms.

The proposed architecture along with the associated partitioning of design/prototyping responsibilities between Rutgers, GA Tech and Lucent is shown in Figure 3.6. The need for a base band and network processor board that would interface to the RF front-end and allow dynamically reconfigurable software and hardware implementations of multiple wireless links supporting individual data rates up to 50 Mb/s and a maximum aggregate data rate of 100 Mb/s. A combination of FPGA for hardware implementation and embedded RISC for software implementation is used. RSIC is preferred to a DSP because of ease of programming.

The tri-band (700 MHz, 2.4 GHz and 5.1 GHz) capabilities using a novel MEMS device from GA Tech provides an important flexibility feature for an experimental platform of this type. The analog front-end would also support two channels, one for measurement and one for data, with bandwidths selectable in 1 MHz increments.

Main Functions of Cognitive Radio System

The main functions of cognitive radios are:

- Spectrum Sensing: Detecting the unused spectrum and sharing it without harmful interference with other users. It is an important requirement of the Cognitive Radio network to sense spectrum holes, detecting primary

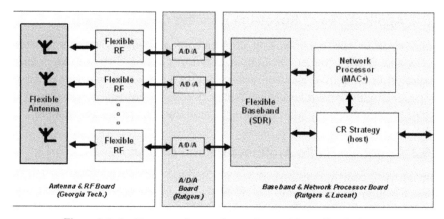

Figure 3.6 Architecture of network-centric cognitive radio platform

users is the most efficient way to detect spectrum holes. Spectrum sensing techniques can be classified into three categories:

1. Transmitter detection: cognitive radios must have the capability to determine if a signal from a primary transmitter is locally present in a certain spectrum, there are several approaches proposed:

   – Matched filter detection
   – Energy detection
   – Cyclostationary feature detection

2. Cooperative detection: This refers to spectrum sensing methods where information from multiple Cognitive radio users are incorporated for primary user detection.

3. Interference based detection.

- Spectrum Management: Capturing the best available spectrum to meet user communication requirements. Cognitive radios should decide on the best spectrum band to meet the quality of service requirements over all available spectrum bands, therefore spectrum management functions are required for Cognitive radios, these management functions can be classified as Spectrum analysis and Spectrum decision.

- Spectrum Mobility: This is defined as the process when a cognitive radio user exchanges its frequency of operation. Cognitive radio networks target to use the spectrum in a dynamic manner by allowing the radio terminals to operate in the best available frequency band, maintaining

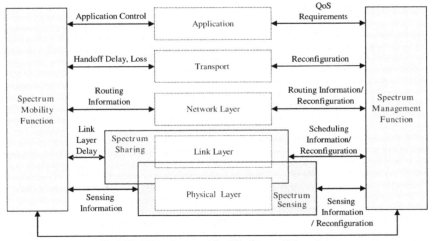

Figure 3.7 Cognitive radio network communication functionalities (adapted from [5])

seamless communication requirements during the transition to better spectrum.

- Spectrum Sharing: Providing the fair spectrum scheduling method, one of the major challenges in open spectrum usage is the spectrum sharing. It can be regarded to be similar to generic media access control MAC problems in existing systems.

The Cognitive Radio Network communication functionalities are shown in Figure 3.7.

3.6.4 Benefits of Cognitive Radio

- Autonomously exploit locally unused spectrum to provide new paths to spectrum access.
- Cognitive enabler for seamless services in heterogeneous network environment with a single terminal.
- Flexible spectrum usage, an increased awareness in network nodes, and more autonomous decisions by communication devices.
- Avoiding interference with licensed or unlicensed users.
- Ideal SDRs will have a dramatic improvement in overall system performance relative to the initial generations SDRs.

3.7 Evolutions of ITU-R Standards

ITU-R has started the process of standardization for IMT Advanced. The root term IMT now encompasses IMT-2000 and IMT Advanced. ITU-R has published requirements that are to be fulfilled by candidate radio access technologies before they can be taken up for evaluation and subsequent inclusion in the ITU-R suite of standards for IMT Advanced.

To come up with a standard that can meet and exceed these requirements, 3GPP has started work on Advanced LTE to improve upon the capabilities of LTE. 3GPP has published the requirements to be met by a candidate Radio Access Technology (RAT) to qualify for evaluation for Advanced-LTE. In [8, 9] the comparative requirements set forth between 3GPP for LTE with those for LTE-Advanced are provided.

It is evident that at least a two to three times increase in throughput and spectral efficiency is required, along with a substantial increase in data rates. To achieve these spectral efficiencies and data rates, a variety of techniques are necessary, some of which are explained in the subsequent sections.

3.8 Femto Cell Technologies

Since the link performance (bps/Hz/link) of current 3G standards is very close to the Shannon limit, recent research has been focused on trying to optimize the layout of the cellular structure so that the area spectral efficiency (bps/Hz/km^2 of area) can be improved. One way of doing this is to make the cells as small as possible. This can lead to problems in cases of mobile users. However, it has been observed that more than 50% of the calls made over cellular networks are made by indoor users, who are either stationary or move slowly. This observation has led to the belief that if a small base station (called a Femto cell) can provide coverage for indoor users, then it would massively reduce the load on the outdoor base station (BTS). In this case, the outdoor BTS would only cater to mobile users. It is expected that the overall bps/Hz/km^2 of area would increase.

A Femto cell is a small cellular base station, typically designed for use in residential or small business environments. It connects to the service provider's network via broadband such as DSL or cable. A Femto cell thus allows service providers to extend service coverage indoors, especially where access would otherwise be limited or unavailable. The Femto cellFemto cell incorporates the functionality of a typical base station but extends it to allow a simpler, self-contained deployment. A generic architecture for a typical

Femto cell deployment is shown in Figure 3.8. The Femto cell has a smaller coverage, less capacity, and consumes much less power compared to a Macro BTS. Thus thousands of Femtos can co-exist in an area covered by a single Macro cell.

The backhaul can be through Ethernet, ADSL or Cable, depending on the broadband connection of the subscriber. The user can use the legacy 2G/3G handsets while communicating with the Femto cell. Capacity is typically limited to four to eight simultaneous voice/data sessions, with coverage up to 100 meters. Most Femtos can be installed as plug-and-play devices. Seamless handoffs with the Macro network provide ubiquitous network connectivity for the user with optimum quality and coverage.

3.8.1 Advantages of Femto Cells

- Improvement in indoor coverage and data rates.
- CAPEX and OPEX reduction for the service provider.
- Seamless connectivity for subscribers due to use of legacy handsets.
- Can be used to network the home devices together.

3.8.2 Major Challenges in Femto CellDeployment

However, Femto cell technology is not without its challenges, and the way the R&D organizations and industry responds to these challenges will largely determine the success of Femto cells. Some of these challenges are:

- Allocation of spectrum between Femto cells and outdoor BTSes.
- Management of interference between Femto cells and outdoor BTSes.
- The legacy core network generally connects to only a few thousand outdoor BTSes. It will now be expected to connect to millions of Femto cells.
- Provisioning of QoS for the millions of Femto cells.
- Exploration of cheaper time and frequency synchronization options at the Femto cell.
- Providing secure communication over the Femto cell.

3.8.3 Femto Cell Standardization

All the leading telecom equipment manufacturers and standardization bodies are pursuing Femto cell technology with interest. The Femto forum was formed in 2007 to co-ordinate these efforts. The goal of the Femto forum is to

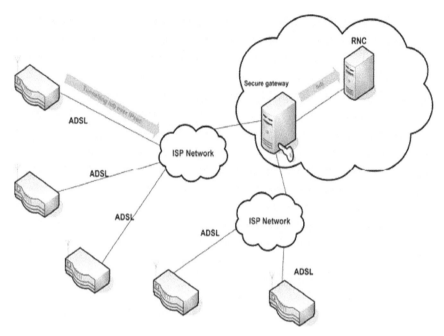

Figure 3.8 A typical Femto cell network topology (adapted from [7])

identify and determine regulatory requirements, network and interoperability issues and radio and physical layers. The two major standardization bodies for cellular and PCS wireless standards 3GPP and 3GPP2, have also been working on specifications related to Femto cells. Technical specifications were recently released by 3GPP in Release 8 for HNBs based on WCDMA and LTE standards. Similarly, 3GPP2 is on course to publish Femto cell specifications for CDMA2000 and EV-DO standards in the last quarter of 2009.

3GPP Specifications for Femto Cells
In UMTS a Femto cell is called a Home Node B (HNB), where access to services may be provided via a UTRAN cellular base station which uses standard Uu interface between UE and HNB. In LTE Femto cell is called HeNB, where access to services may be provided via an E-UTRAN cellular base station that uses standard LTE-Uu interface between the UE and the HeNB.

VPLMN HPLMN

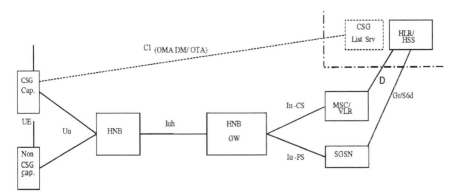

Figure 3.9 Generic architecture of HNB

H(e)NBs provides attractive services and data rates in indoor environments. Conventional UTRAN architecture is not optimally suited for this as it was developed and defined under the assumption of coordinated network deployment where as H(e)NBs are typically associated with uncoordinated and large scale deployment. A generic architecture for a HNB deployment is shown in Figure 3.9.

The HNB GW is specific to the Home Node B architecture and aggregates the traffic from millions of HNBs to feed it into the traditional packet and circuit switched core networks. Communication between the HNBHNB and the HNBHNB GW is secured by a mandatory Security Gateway (SeGW) function, which is not shown in the figure. The SeGW may be implemented either as a separate physical entity or integrated into the HNB GW. The C1 (OMA DM/OTA) interface shown in the above figure is optional and it is used to update allowed Closed Subscriber Group (CSG) lists on CSG-capable UEs.

A generic architecture for a HeNB deployment has three variations as per 3GPP standards on H(e)NBs:

- The first variant uses HeNB deployment with dedicated HeNB GW, as shown in Figure 3.10.
- The second variant uses HeNB deployment without HeNB GW.
- The third variant uses HeNB deployment with HeNB GW for control plane.

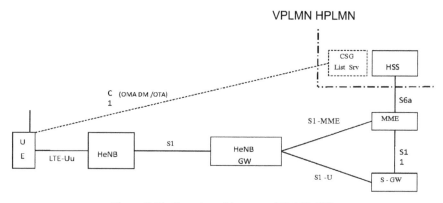

Figure 3.10 Generic architecture of HeNB GW

In order to minimize the impact on the existing overall network, the H(e)NB operates with legacy terminals and core network and also minimizes the impact on existing (E)UTRAN interfaces.

3.9 Conclusions

This chapter has discussed various requirements and challenges for future radio technologies. Future radio access technologies are required to focus on providing higher data rates, new frequency bands, flexible spectrum usage and energy efficiency. Cognitive radios are environment aware, self reasoning and learning capable radios that can change the parameters or protocols depending on the environment in which they operate. Hence cognitive radio is very important technology for the future. This document has also discussed the evolution of ITU-R standards, the improvement in spectral efficiency by Femto cell technology, the 3GPP specification for Femto cell and the comparison of LTE and Advanced LTE requirements. The Advanced LTE has better capabilities than LTE.

3.10 Future Challenges

- Development of new dynamic spectrum access and management techniques.
- Future spectrum access in presence of cognitive radios will introduce more challenges.

- Spectrum sensing, spectrum analysis and decision making in cognitive radio context will be the major challenge.
- Development of technologies which will provide higher data rates, lower latencies and high QoS.
- Study on deciding which bands would be available for initial Cognitive Radio trials.
- Develop Cognitive Radio algorithms to enable co-existence between IEEE 802.11b and 802.16a networks in the same unlicensed band.
- Develop two wireless systems share radio resources in frequency, space and time, and coordination methods to be used to reduce the mutual interference.
- The effect of interference produced by the man-made/natural environment.
- Study on defining "harmful interference" within the authorized Cognitive Radio bands.
- Develop practical algorithms for the spectrum-sensing Cognitive Radio, for exchanging spectrum monitoring data between cooperating communications nodes.
- Study on defining the rights of the licensed spectrum band users where Cognitive Radio technology is authorized [6].

References

[1] P. Liu, Z. Tao, Z. Lin, E. Erkip, and S. Panwa, Cooperative wireless communications : A cross-layer approach. *IEEE Wireless Communications*, 13(4):84–92, August 2006.

[2] A. Nosratinia, T.E. Hunter, and A. Hedayat, Cooperative communication in wireless networks (adaptive antennas and MIMO systems for wireless communications). *IEEE Communications Magazine*, 42(10):74–80, October 2004.

[3] FCC, ET Docket No. 03-222 Notice of proposed rulemaking and order, December 2003.

[4] J. Mitola III, Cognitive radio: An integrated agent architecture for software defined radio. PhD Thesis, KTH Royal Institute of Technology, Stockholm, Sweden, 2000.

[5] I.F. Akyildiz, W.-Y. Lee, M.C. Vuran, and S. Mohanty, Next generation/dynamic spectrum access/cognitive radio wireless networks: A survey, *Science Direct Computer Networks Journal*, 50:2127–2159, September 2006.

[6] K. Sridhara, A. Chandra, and P.S.M. Tripathi, Spectrum challenges and solutions by cognitive radio: An overview. *Wireless Personal Communications*, 45(3):281–291, May 2008.

[7] The case for home base stations, Pico Chip Technical White Paper, v1.1, April 2007.

[8] 3GPP TR 25.913, Requirements for E-UTRA and E-UTRAN.

[9] 3GPP TR 36.913, Requirements for Further Advancements for E-UTRA.

[10] V. Chandrasekhar, J.G. Andrews, and A. Gatherer, Femto cell networks: A survey. *IEEE Communications Magazine*, 46(9):59–67, September 2008.
[11] L. Sabel, Software-defined radio: The solution for multi-standard multimedia in the mobile environment. *EBU Technical Review*, 4/8, January 2007.

Biographies

Bhuwansing A. Damahe has a B.E degree in Electronics and Power and an M.E. degree in Electronics with a specialization in computing. He has worked with various organizations in various capacities, like PCS Data Products Ltd, Bombay as Senior Hardware Engineer, Mina Fax Electronics System Pvt. Ltd, Ahmedabad as Senior Engineer, College of Applied Science, Aurangabad, as Senior Lecturer, DOEACC Center, Aurangabad, as Senior Lecturer. Currently, he is working with L&T Institute of Technology, Mumbai as the Principal. He has published six papers at national level conferences and one paper at an international conference. He is a member of a number of professional and advisory bodies such as Board of studies on SSC, Government of Maharashtra, various syllabus setting committees for IT in Pune and Marathawada University, etc. He is a life member of Indian Society for Technical Education (ISTE), life member of Institute of Telecommunication and electronics Engineers (IETE), life member of Computer Society of India (CSI), and member of IEEE. He has traveled abroad on a number of occasions. He has worked with Professor P. Rebaud at Ecole Polytechnique Federale de Lausanne Switzerland (EPFL) on a project entitled "Digital interfacing of capacitive pressure micro sensor" during May 1995.

Olive Jesudas is currently working as a faculty member with the L&T Institute of Technology, Mumbai, the last five years in the Department of Electronics and Communication Engineering. Previously, she worked in the Regional Engineering College, Trichy for four years. Apart from teaching, she has contributed in other key administrative and departmental roles such as organizing inter-departmental competitions, ISO auditing, project coordination, mentoring, etc. She has also conducted various short-term training programmes, and participated in the panel in various national level and state level conferences. To enhance industry-institute interactions, she has taken part in a six-week industrial training in ZigBee at L&T Embedded Systems and applications. She has attended the International Workshop on Global Information Communication Technology (ICT) standardization

Forum for India at Lonavala, India in March 2009. She has a B.E in Electronics and Communication Engineering and an M.E in Communication Systems from REC Trichy.

Girish Manglani has six years experience in the telecommunications field. He graduated with an M.E. degree in Information Systems from the IIT of Kanpur in 2003. He is currently working with Reliance Communications as a Senior Manager. His areas of interest include various physical layer aspects of multiple access technologies like CDMA and OFDMA.

4

Convergence and Next Generation Networks

Jaydip Sen[1], Munir Sayyad[2] and Basavaraj Hooli[3]

[1]*TCS, Kolkata, India*
[2]*Reliance Communications, Mumbai, India*
[3]*Mphasis, Pune, India*

Communication networks have become a key economic and social infrastructure in many countries in the world. The telecommunication network infrastructure is crucial to the national and international exchange of goods and services, and acts as a main catalyst in changing economic interrelationships through rapid technological change and the proliferation of a range of new services. With the development of the Internet, the role of communication networks has evolved and their importance increased. The advent of higher access speeds, in many cases symmetric speeds, available to business and to residential subscribers, has also increased the role of communication infrastructures by expanding the available range of services. High speed networks are increasingly helping to resolve ongoing societal concerns in areas such as the environment, healthcare and education, and are increasingly playing a role in social networking. However, for the potential of new network technologies to be realized, the market will require that these networks have universal, or close to universal coverage. The full potential of networks is only likely to be achieved where markets are effectively competitive and solutions have adequate coverage to most geographic areas.

Technological innovation, stimulated through digitalization, has been a major factor in driving change in the communications market. This innovation is reducing costs and enhancing the capability of networks to support new services and applications. A key innovation which is expected to bring further

R. Prasad (ed.), Future Trends and Challenges for ICT Standardization, 107–192.
© 2010 River Publishers. All rights reserved.

significant changes in the communications market is the transformation from circuit-based public switched telecommunication networks to packet-based networks using the Internet Protocol, so-called *next generation networks* (NGN). NGN is expected to completely reshape the present structure of communication systems and access to the Internet. The present structure of vertically independent, although interconnected, networks may be transformed into a horizontal structure of networks based on Internet Protocol. Investment requirements for NGN are high and, as for any investment, there are risks. Policies need to ensure that risks and uncertain returns are compensated while ensuring competition since, without competition, the benefits of high speed broadband and NGN will not be realized.

The communications sector is undergoing significant changes, with the emergence of a number of platforms available to provide a different range of services. Some of these platforms are complementary to each other, while others are competitive, or can provide a valid substitute for some of the services provided. Up till now, the most important communications platform in most of the developing countries has been the public switched telecommunication network (PSTN) which provides access to all households and buildings. This universality in providing access has also meant that the network has generally been designated as one for universal service.

This chapter focuses on the area where the most significant changes are taking place in the communication sector. The objective is neither to give an overview of all communication platforms, nor is it aimed to assess the relative extent to which different platforms complement or compete with each other. The central theme of the chapter is to examine the developments in what is commonly referred to as next generation access networks and next generation core networks and their role in convergence. The focus on the next generation access and core networks is because of their huge impact and change these network will bring in telecommunication sector.

The developments in new communication structures and the impetus they are expected to give to the present process of *convergence* in networks, services and terminals are expected to lead also to new policy challenges. Convergence, by changing service boundaries, service characteristics and stimulating the offer of new services, may require that new markets are regulated differently than existing ones. It remains to be seen to what extent the deployment of NGN and convergence will facilitate the process of creating durable competitive conditions in communication markets or will raise further obstacles to the creation of competition. It is fairly evident, however, that changes taking place as a result of investment in next generation access and core networks and the convergence of technologies, services and markets will require reviews and rethinking of existing policy and regulatory frameworks.

The rest of the chapter is organized as follows. Section 4.1 introduces the concept of convergence and NGN. Section 4.2 presents details of the access networks and core networks for the next generation converged applications and also describes the drivers of the NGN. Section 4.3 describes various possible interconnection frameworks for NGN deployment. Section 4.4 presents broadcasting convergence in all-IP networks. Section 4.5 discusses the security aspects of the NGN. Section 4.6 introduces IP multimedia subsystem (IMS) and presents its architectural details. Section 4.7 describes a case study based on IMS in residential networking environment. Section 4.8 discusses the details of various standard bodies involved in the activities related to convergence and NGN. Section 4.9 presents the current and future trends of the Indian telecommunication industry with particular attention on the challenges the industry will face in deployment of NGN and converged services. Section 4.10 presents some open research problems in convergence and finally Section 4.11 concludes the chapter.

4.1 Convergence and NGN

Convergence in network technologies, services and in terminal equipment is at the basis of change in innovative offers and new business models in the communications sector. The utilization of the term "convergence" represents the shift from the traditional "vertical silos" architecture, i.e. a situation in which different services were provided through separate networks (mobile, fixed, CATV, IP), to a situation in which communication services will be accessed and used seamlessly across different networks and provided over multiple platforms, in an interactive way. Already in the 1990s the possible impact of digitalization and convergence between telecommunications and broadcasting was under examination and proposals made for changes in existing regulation. The growing role of the Internet in the economy and society has enhanced the process of convergence and its rate of change.

What is Convergence?

The path towards convergence was led mainly by the increasing digitalization of content, the shift towards IP-based networks, the diffusion of high-speed broadband access, and the availability of multimedia communication and computing devices. Convergence is taking place at different levels.

Network convergence driven by the shift towards IP-based broadband networks. It includes fixed-mobile convergence and "three-screen convergence" (mobile, TV and computer).

Service convergence stems from network convergence and innovative handsets, which allows the access to web-based applications, and the provision of traditional and new value-added services from a multiplicity of devices.

Industry/market convergence brings together in the same field industries such as information technology, telecommunication, and media, formerly operating in separate markets.

Legislative, institutional and regulatory convergence or at least co-operation is taking place between broadcasting and telecommunication regulation. Policy makers are considering converged regulation to address content or services independently from the networks over which they are provided (technology neutral regulation).

Device convergence: most devices include today a microprocessor, a screen, storage, input device and some kind of network connection; increasingly they provide multiple communication functions and applications.

Converged user experience: unique interface between end-users and telecommunications, new media, and computer technologies.

The process towards convergence has been based on an evolution of technologies and business models, rather than a revolution. This process has led to:

- Entry of new players into the market.
- Encasing competition among players operating in different markets.
- The necessity for traditional operators to co-operate with companies previously in other fields.

As a result, convergence touches not only the telecommunication sector, but involves a wider range of activities at different levels, including the manufacturer of terminal equipment, software developers, media content providers, ISPs, etc.

Previously distinct communication networks and services are today converging onto one network, thanks to the digitalization of content, the emergence of IP, and the adoption of high-speed broadband. Traditional services such as voice and video are increasingly delivered over IP networks and the development of new platforms is facilitating the provision of converged services (Table 4.1). These converged services are appearing in markets as "triple" or "quadruple" play offers which provide data, television, fixed and mobile voice services.

The process of convergence has also been facilitated by the opening up of telecommunication markets to competition. Although large telecommunication operators have played a role in the process of convergence, new market players have moved rapidly, and often in an unpredictable way, adopting different market models from traditional telecommunication firms. Voice over IP is a clear example of such services, disrupting traditional markets, pushing towards adoption of next generation networks and facilitating convergence. Internet service providers started offering VoIP as a cheaper way to communicate over the Internet. Services were offered on a "best-effort" basis by third parties, over any Internet connection. Today the market for VoIP services is varied, with network access operators providing VoIP as a replacement for PSTN voice telephony, often guaranteeing access to emergency services, or a certain QoS. Internet service providers continue to offer access to VoIP services from multiple platforms and from anywhere in the world. Mobile VoIP is also emerging, both as a service provided by the network operator or as an application that can be downloaded on any Wi-Fi enabled handset. Initiatives, such as Google's "Android", are likely to put pressure on existing mobile operators to charge flat rates for mobile Internet access, thus eventually increasing the degree of substitutability between mobile and fixed Internet access (in terms of price rather than speed) [1].

On the content side increasing competition is taking place between network access operators, including wireless, cable or satellites, all offering video, music, or other content to their users. A growing number of operators are also focusing on mobile content, in particular on the possibility to download music, or access applications and online services from a mobile device. The possibility to provide video content over IP is often seen as a new way to propose content to users, and as an opportunity for network operators to enlarge the range of services they offer to their customers. Content services, especially those over managed IP networks, have still not exploited their full potential. In most cases, access to content is offered in a form very similar to traditional broadcasting, with defined timetables, geographical distribution, rigid copyright schemes, a very low degree of interactivity, and a traditional billing scheme although a number of operators are now beginning to offer more flexible programming with video on demand and distribution of video content from popular Internet sites. Changes are often taking place as a result of an increasing number of users creating and exchanging their own content on a multiplicity of devices, which imply a shift away from simple pass-

Table 4.1 An all IP-based converged environment

Telecommunication environment	Next-generation converged environment
Single purpose networks	Multi-purpose networks
PSTN, cellular, broadcast	IP network (providing voice, video and mobile services)
Narrow-band	Broad-band
Vertical silos	Destroys compartmentalization Traditional boundaries between industry segments (e.g., telephony, cable TV, broadcasting, wireless) are blurring-need to re-think market definitions (product definition and geographic boundaries definition)
Network-service link	New services and content developed independently of the network
Operators control services to end users	Increased consumer control

ive consumption of broadcasting and other mass distribution models towards more active choosing, interacting, the creation of content, and the emergence of a participatory culture [2]. These developments also increase the need to communicate, and the demand for symmetric communications.

With the growing offers from access platforms, and the different types of video services and applications available – digital terrestrial, IPTV, HDTV, Video on Demand, but also disruptive applications such as Joost [3] or Sling box [4] – the concept of "social value" of terrestrial broadcasting may become weaker, while the impact of these new services remains to be assessed.

Although the extent and effects of such convergence are yet to be seen, the phenomenon is already challenging the existing remit of many sector – specific domestic regulations. For example, the impact of convergence on competition is likely to be mixed. On the positive side, the move towards next generation networks, able to deliver a wide range of communication services, creates a schism in many traditional market definitions. While in the past telecommunication companies only offered fixed-line voice, and policy makers could easily define the specific market and make regulatory decisions, today the convergence of video, voice and data on next generation broadband networks can lead to more competition in individual markets for each of these services. As a result, convergence touches the telecommunication, cable television and broadcasting sectors, and involves a wider range of activities at different levels, going from manufacturers of terminal equipment, software developers, media content providers, ISPs, etc.

Bundled Services

Bundling refers to the sale of a number of services combined in a single price package, usually excluding the possibility that customers can obtain a single service without taking or paying for the other services in the bundle.

Bundling of services can help generate economies for the supplier through, for example, reduction in service marketing charges, customer acquisition costs, billing charges, etc. For the client bundling often has the advantage that prices are lower compared to having to subscribe to each service individually, however, customers may not want all the services offered in a bundle. A client who does not want IPTV may be obliged to pay for these services when subscribing to certain triple play bundles.

At the same time a bundle, while normally offering a better price than purchasing the same services separately, is also difficult to assess when trying to compare prices across a range of different offers. A service provider may also use a service in the bundle to cross-subside other services using this to obtain an unfair market advantage. Bundling may also make it difficult for regulators to define markets, assess market power, and therefore understand whether or not dominance exists in a given market.

On the other hand, the trend towards horizontal integration of markets and services could lead to strengthening of market power, as there may be relatively few companies in a country that can provide a combined video, voice and data offering. This may lead to a reduction in competition for the communications sector as a whole. In addition, bundling of services may make it more difficult to determine the extent to which prices are cost-oriented, allowing cross-subsidization between services. Service convergence and the shift towards next generation networks could therefore contribute to the creation of additional bottlenecks and control points, which may need to be addressed by the regulator.

In this context, next generation networks (NGN) provide the technical underpinning of convergence, representing a single transport platform on which the carriage of previously distinct service types (video, voice, and data) "converges", together with new and emerging services and applications. While different services converge at the level of digital transmission, the separation of distinct network layers (transport, control, service and applications functions – Figure 4.1) provides support for competition and innovation at each horizontal level in the NGN structure. At the same time NGNs also create strong commercial incentives for network operators to bundle, and therefore increase vertical and horizontal integration, leveraging their market power across these layers. This may bring about the need for closer regulatory and policy monitoring, in order to prevent the restriction of potential development of competition and innovation in a next generation environment, and therefore the risk of reducing benefits for consumers and the potential of new networks for economic growth.

4.2 Next-Generation Networks

Although there is a significant amount of work underway in standardization forums on NGN, at the policy level, there is still no complete agreement on a specific definition of "NGNs". The term is generally used to depict the shift to higher network speeds using broadband, the migration from the PSTN to an IP-network, and a greater integration of services on a single network, and often is representative of a vision and a market concept. From a more technical point of view, NGN is defined by the International Telecommunication Union (ITU) as a "packet based network able to provide services including telecommunication services and able to make use of multiple broadband, QoS enabled transport technologies and in which service related functions are independent from underlying transport-related technologies". NGN offers access by users to different service providers, and supports "generalized mobility which will allow consistent and ubiquitous provision of services to users" [5]. NGN, also defined as "broadband managed IP networks", includes next generation "core" networks, which evolve towards a converged IP infrastructure capable of carrying a multitude of services, such as voice, video and data services, and next generation "access" networks, i.e. the development of high-speed local loop networks that will guarantee the delivery of innovative services.

4.2.1 Next-Generation Access Networks

The definition of next generation access networks is usually specific to investment in fiber in the local loop, i.e. fiber replacing copper local loops, able to deliver next generation access services – i.e. an array of innovative services, including those requiring high bandwidth (voice, high-speed data, TV and video). However, while next generation access networks tend to refer to a specific technological deployment, there are other technologies which can compete in providing some of the services which it is envisaged will be provided by NGNs. There are also other technologies which may not be able to fully compete with NGN access networks in terms of capacity and the plethora of bundled offers which NGNs can provide, but may be perfectly suitable for users who do not have the need for high capacity access. The different technologies available include existing copper networks upgraded to DSL, coaxial cable networks, powerline communications, high speed wireless networks, or hybrid deployments of these technologies. Although fiber, in particular point-to-point fiber development, is often described as the most "future proof" of network technologies to deliver next generation access [6], there are likely to be a number of alternative and complementary options for deployment of access infrastructures by incumbent telecommunications operators, and new entrants.

Cable Television

(CATV) operators have begun to upgrade their infrastructure to hybrid fiber copper (HFC) allowing for bidirectional traffic and using Docsis [7] technology to increase network capacity. These developments are allowing CATV companies to offer voice and Internet access (data services) in competition with telecommunication companies which through their offer of Internet TV have begun to compete with CATV companies. Offering data and voice services, in addition to television, helps cable companies differentiate their product offering from satellite providers. The bandwidth provided by cable networks, using Docsis 3.0, will allow for 160 Mbit/s downstream and 120 Mbit/s upstream for end-users.

Broadband Wireless Access

Broadband wireless access (BWA) technologies aim, providing high speed wireless access over a wide area. Certain early fixed wireless access technologies, such as local multipoint distribution service (LMDS) and multi-channel multipoint distribution service (MMDS), never gained widespread market adoption. WiMAX technologies, the IEEE 802.16 set of standards that are the foundation of WiMAX certification, and similar wireless broadband technologies, are expected to address some of these shortcomings, and fill market gaps left by wired networks, or compete with wired access providers. The WiMAX Forum has estimated that new WiMAX equipment will be capable of sending high-speed data over long distances (a theoretical 40 Mbit/s over 3 to 10 kilometers, in a line-of-sight fixed environment). When multiple users are connected, capacity sharing will significantly reduce speeds for individual users sharing the same resource [8].

Wi-Fi (or wireless fidelity) refers to wireless local area networks that use one of several standards in the 802.11 family. Wi-Fi allows LANs to be deployed without cabling for client devices, typically reducing the costs of network deployment and expansion. Due to its affordability, scalability and versatility, its popularity has spread to rural and urban area. Wi-Fi range is usually limited to about 45 m indoor and 90 m outdoors, however Wi-Fi technologies can also be configured into point-to-point and point-to-multipoint networks in order to improve their range and provide last mile fixed wireless broadband access. One way to serve remote areas which cannot be reached with the above-mentioned technologies is with wireless "mesh" solutions. They often include a satellite backhaul connection through Very Small Aperture Terminals (VSAT), usually coupled with wireless technologies such as Wi-Fi. This combination allows access to telecommunication and data services even to more remote areas, albeit with limited (and expensive) bandwidth [9].

Terrestrial wireless services offer the opportunity to deploy competing access infrastructure. However, they may offer different service characteristics to fixed-line services in terms of coverage, symmetry and speeds. These networks may be less suitable to deliver sustained high bandwidth connections for larger numbers of users, or for high bandwidth applications, such as High Definition TV on demand. In addition, wireless service deployments are constrained by spectrum availability. At the same time, the economics of their deployment is often relatively scalable, which means that they have lower economic barriers to entry compared to fiber deployments [10]. While they may not be a complete substitute, they can complement wireline networks for specific services [11].

Broadband over Powerlines (BPL)

Use of the power grid as a communications network, or "powerline communications" appears to provide a series of advantages, offering not only voice, but also broadband services, with the connection speed not dependent on distance from the telephone exchange (as happens with DSL) or number of customers connected (as with cable). With this system a computer (or any other device) would need only to plug a BPL "modem" into any outlet in an equipped building to have high-speed Internet access. Notwithstanding the benefits that the availability of an extensive infrastructure can allow, for the moment the service provision is far from standardized [12], and the capacity of bandwidth provided through BPL is still being questioned.

3G Mobile Networks

The term NGN frequently encompasses some kind of fixed-mobile convergence (FMC) [13], as it allows the transition from separate network infrastructures into a unified network for electronic communications based on IP, which facilitates affordable multiple business models, seamlessly integrating voice, data and video. The introduction of 3G technology supports the transmission of high-speed data with speeds theoretically reaching 2–4 Mbit/s. The operators are expanding their 3G networks across different countries. This will provide higher data speeds to users enabling them to access innovative networks dedicated to provide mobile video or television programming. However, existing 3G technologies will need to be upgraded in order to support very high bandwidth or extensive concurrent usage that may be demanded by users in the future. The future evolution of mobile networks for example using LTE technology (Long Term Evolution) – a next generation mobile technology – may significantly increase speeds, enabling high peak data rates of 100 Mbit/s downlink and 50 Mbit/s uplink.

Satellite Networks

Satellite services are typically dedicated to direct-to-home television and video services, satellite radio, and specialized mobile telephony uses. More recent technological advances such as spot beam technology and data compression algorithms have enhanced the technical efficiency in satellite communications, enabling more efficient use of spectrum, and reduced redundancy. It has resulted in increased effective data density and reduced required transmission bandwidth [14]. Satellite broadband is usually provided to the customer via geosynchronous satellite. Ground-based infrastructure includes remote equipment consisting of a small antenna and an indoor unit. Gateways connect the satellite network to the terrestrial network. Except for gateway locations, satellite broadband is independent of terrestrial infrastructure such as conduits and towers, allowing it to provide coverage also to remote areas.

4.2.2 Next-Generation Core Networks

The next generation *core* networks are defined on the basis of their underlying technological "components" that include – as mentioned in the ITU definition – packet-based networks, with the service layer separated by the transport layer, which transforms them into a platform of converged infrastructure for a range of previously distinct networks and related services. These features may have an impact on traditional business models and market structure, as well as on regulation:

- *IP-based networks*: "Next generation core networks" generally cover the migration from multiple legacy core networks to IP-based networks for the provision of all services. This means that all information is transmitted via packets. Packets can take different routes to the same destination, and therefore do not require the establishment of an end-to-end dedicated path as is the case for PSTN-based communications.
- *Packet-based, multi-purpose*: While traditionally separate networks are used to provide voice, data and video applications, each requiring separate access devices, with NGN different kinds of applications can be transformed into packets, labeled accordingly and delivered simultaneously over a number of different transport technologies, allowing a shift from single-purpose networks (one network, one service), to multi-purpose networks (one network, many services). Interworking between the NGN and existing networks such as PSTN, ISDN, cable, and mobile networks can be provided by means of media gateways.
- *Separation of transport and service layers*: This constitutes the key common factor between NGN and convergence, bringing about the radical change in relationship between network "layers" (transport infrastructure, transport services and control, content services and applications). In next generation networks service-related functions are independent from underlying transport-related technologies (Figure 4.1). The uncoupling of applications and networks allow applications to be defined directly at the service level and provided seamlessly over different platforms, allowing for market entry by multiple service providers on a non-discriminatory basis.

These features may foster the development and provision of new services and constitute a new opportunity for innovation, allowing different market players to create value at the separate functional levels of access, transport, control and services.

However, while initially it was a common assumption that this layered structure would lead to a market model where services could be increasingly provided across the value chain, in a more decentralized manner, today it appears that the network provider will decide whether the "horizontal" model will prevail, or whether they will simply (commercially) vertically integrate transport and services across functional levels, offering bundled services [15].

Currently, bundling of a variety of services is a key trend in the sector, bringing greater competition between formerly distinct sectors. Bundles include all sorts of combinations of fixed and mobile voice calls, Internet

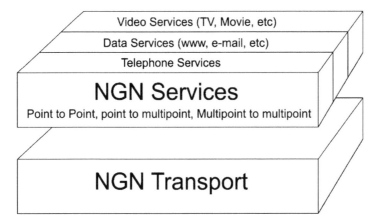

Figure 4.1 Separation of functional planes. (adapted from [115])

access and media/entertainment services (Section 4.5). With services and transport commercially integrated at the vertical level, customers are some-how "locked-in" in a vertical relationship with a single operator. This is not negative in itself, as packages are often more convenient, or easier to use, at the same time it is important to maintain the possibility for users to choose which services they want to purchase, and to have clear information about the cost and characteristics of these. The risk would be to create a situation in which the network provider may limit the possibility of users to access IP-based services and applications provided by third parties.

Considering the economic drivers behind the shift towards next genera-tion networks, there is an incentive for the network provider to also become an integrated market player, in order to maintain/extend their user base or benefit from a privileged relationship with subscribers. This raises questions regarding obligations for access to networks by service providers and issues of traffic prioritization [16]. In this context access plays an important role for all service providers to be able to provide their content, services and applications to end users.

One essential feature of next generation networks is the capability to support "generalized mobility which will allow consistent and ubiquitous provision of services to users" [17]. Although core next generation networks tend to be on a fixed infrastructure, the possibility to improve interconnection with mobile networks is being explored, and standardization organizations as well as operator and manufacturers associations are working to the de-velopment of appropriate standards. In addition, the deployment of wireless infrastructures facilitates access to IP networks, and the adoption of increasingly sophisticated devices and handsets will allow an easy access to IP services from anywhere.

The migration process towards IP-NGN potentially entails several structural changes in the core network topology, such as the rearrangement of core network nodes and changes in the number of network hierarchy levels. As a result, an overall reduction in the number of points of interconnection will take place, especially with regard to interconnection points at the lowest level. This could negatively affect alternative operators whose previous interconnection investment may become stranded [18]. For example, BT today has some 1200 exchanges at which competitors have installed DSLAMs, using local loop unbundling to provide broadband and bundled services. In addition, BT has over 700 exchanges at which competitors can connect their voice services. The number and location of points at which competitors could connect their networks to BT's voice services is expected to reduce substantially to at most 108 Metro-node sites, and probably to a subset of these which could number as few as 29, while the number and location of exchanges at which local loop unbundling is likely to be possible are not expected to be affected by the roll-out of 21CN [19].

4.2.3 NGN Drivers and Impact

NGN is an evolutionary process and it can be expected that operators will take different migratory paths, switching to NGN while gradually phasing out existing circuit networks, or building a fully-IP enabled network from the outset [20]. The investment in developing NGN is motivated by several factors (Table 4.2). Telecommunication operators across most of the countries in the world have been faced with a decline in the number of fixed-line telephone subscribers, coupled with a decrease in *average revenue per user* (ARPU), as a result of competition from mobile and broadband services [21]. Traditional sources of revenue (voice communications) have declined rapidly and fixed-lines operators are subject to an increase in competitive pressure in the market to lower tariffs and offer innovative services. This has generated pressure from the investors' community to decrease the cost and complexity of managing multiple legacy networks, by disinvesting from non-core assets and reducing operational and capital expenses.

In this context, the migration from separate network infrastructures to next generation *core* networks is a logical evolution, allowing operators to open up the development of new offers of innovative content and interactive, integrated services, with the objective to retain the user base, attract new users, and increase ARPU. NGN is therefore often considered essential for network operators to be "more than bit pipes" [22] and to strategically position themselves to compete in the increasingly converged world of services and content, where voice is no longer the main source of revenue, and may become a simple commodity. The investment in next generation access networks – both wired and wireless – will be necessary in order to support the new services enabled by the IP-based environment, and to provide increased quality. At the same time, the important investment necessary to develop next generation infrastructures brings about new economic and regulatory issues, which will be analyzed in the following sections.

Although the shift in the migration to all-IP networks is taking place at different paces in different countries, several operators have already updated their transport networks, and are now dealing with NGN at the local access level. Solutions embraced by fixed operators may also increasingly support IP Multimedia Subsystem (IMS), to enable fixed-mobile convergence [24].

For the moment the most common services provided through the new networks are the provision of PSTN/ISDN emulation services, i.e. the provision of PSTN/ISDN service capabilities and interfaces using adaptation to an IP infrastructure, and video on demand (VoDs). At the same time the business world is showing an increasing interest in new NGN-enabled services and applications. Companies are migrating their Time Division Multiplexing switches to IP in order to enable integrated applications for specific industry-based functionalities and purposes [25].

Progress in the field of mobile (cellular) communications is taking shape with the development of the IMS standard [26]. For the moment two services have been standardized under the IMS protocol, Push to Talk over Cellular (PoC) and Video Sharing [27]. Prominent telecommunication network equipment suppliers are actively supporting the take up of IMS and some of them are implementing IMS strategies and commercial IMS products [28]. IMS is seen as the enabler for the migration to next generation networks of mobile operators and therefore for the implementation of fixed-mobile convergence. No evident killer application has currently emerged, with many operators focusing on one specific service: voice. Facilitating the use of voice applications, enabling users to handle their calls easily between fixed and mobile networks, and to receive calls wherever they are, is fundamental for the take-up of the service. Operating in an IMS environment would allow a seamless handover from WLAN (fixed) to mobile during calls (Voice Call Continuity).

Table 4.2 NGN drivers [23]

Economic Drivers	Technological Drivers	Social Drivers
• Erosion of fixed line voice call revenues • Competitive pressure from new entrants in high-margin sectors of the market (long distance, internal) and vertically integrated operators (triple-play bundles) • Saturation of both fixed and mobile telephone services • Retain and expand users' base, lower customer churn • Ability to expand into new market segments • Possibility of 'ladder of investment', i.e. a phased approach for investment, initially targeting more densely populated areas, and then gradually	• Obsolescence of legacy networks, plus cost and complexity of managing • Multiple legacy networks • Lower capital and operational expenses • Increased centralization of routing, switching and transmission, lower transmission costs over optical networks • IP-based networks • Enable the provision of cheaper VoIP services as a replacement for PSTN voice services • IP-based networks enable the provision of a wider range of services, and allow bundling of services (triple and quadruple play). • Evolution and convergence of terminal	• Demand for innovative, high- bandwidth, services (HDTV, VoIP, etc.) • Demand for more targeted or personalized content (on demand multimedia services, mobility) • Demand for increased interactivity: possibility to interact actively with the service, growing interest for user-created content • Demand for evolved and more flexible forms of communications, including instant messaging, video conferencing, P2P, etc. • Business demand for integrated services, in particular, in case of multi-national structures, which need to link different national branches, guaranteeing a flexible and secure access to centralized resources and intelligence

In order for real-time voice call to be offered seamlessly between the circuit switched domain and the Wireless LAN interworking with IMS architecture, the Third Generation Partnership Project (3GPP) [29] is currently working to develop the appropriate Technical Specifications to define this functionality as a standard 3GPP feature. The study by 3GPP of the standard is underway [30]. In the meantime, fixed-mobile converged services have been launched by some mobile operators with access to fixed networks, using a different standard – Unlicensed Mobile Access (UMA) [31] – allowing users to seamlessly switch from fixed to mobile networks (see Section 4.3.4).

In addition, increasing competitive pressure on mobile carriers is coming from the IP world. Thanks to the availability of dual-use devices and Wi-Fi hotspots, service providers – such as Skype, Google, and others – are able to offer on the market a host of new services for mobile users in a very short period of time. This rapidity constitutes an important comparative advantage, which in some cases provoke the reaction of mobile operators (and manufacturers), tending to limit the services and applications users can access from their mobile handset.

4.2.4 Internet and NGN

Technological developments associated with next generation networks should help combine the characteristics of the traditional telecommunication model, and of the new Internet model, dissolving the current divisions and moving towards a harmonized and coherent approach across different platforms, gradually bringing to full convergence fixed and mobile networks, voice, data services, and broadcasting sectors. In short, in the future the choice of the technology used for the infrastructure or for access will no longer have an impact on the kinds and variety of services that are delivered.

This however does not reflect the current situation, where the two worlds still have different visions and commercial models (Figure 4.2).

The telecommunications tradition emphasizes the benefits of higher capacity local fiber access facilities, and powerful network intelligence. Access in this context should be simple and reliable, with centralized network management and control to guarantee the seamless provision of a wide range of services, bundled network-content-applications offers, and one-stop shop solutions.

On the other hand, the Internet world traditionally focuses on edge innovation and control over network use, user empowerment, freedom to choose and create applications and content, open and unfettered access to networks, content, services and applications. Freedom at the edges is considered more important than superior speed of managed next generation access networks.

Indeed, the "Internet" still represents different things to different people, and next generation networks are seen as both a possibility for improved services or as a way to constrain the Internet into telecommunication boundaries, adding new control layers, capable of discriminating between different content, and "monetize" every single service accessed.

Services provided over next generation networks will differ from services currently provided over the public Internet which is based on a "best effort" approach, where the quality of transmission may vary depending on traffic loading and congestion in the network, while with NGN packet delivery is enhanced with Multi Protocol Label Switching (MPLS). This allows operators to ensure a certain degree of Quality of Service – similar to the more constant quality of circuit switched networks – through traffic prioritization,

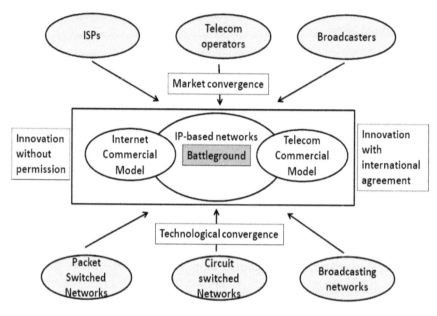

Figure 4.2 The convergence model (adapted from [116])

resource reservation, and other network-based control techniques, as well as to optimize network billing as in circuit-switched transport [32].

The concept of network-based control seems to be the main difference between the public Internet approach and next generation managed IP networks approach. NGN offers the possibility to provide a detailed service control and security from within the network, so that networks are aware of both the services that they are carrying and the users for whom they are carrying them, and are able to respond in different ways to this information. In contrast, the Internet aims to provide basic transmission, remaining unaware of the packets/services supported. While the Internet model remains therefore completely open to users and new applications and services, in managed IP networks operators are able to control the content going through the network [33]. In turn, this may have negative implications for the content of third party providers if their traffic is discriminated against in relation to that of an integrated operator.

In addition, the chapter presents the technical challenges for migration towards the converged next generation networks (NGN). This migration requires changes in the network topology which potentially involves several structural changes, such as a re-organization of core network nodes and changes in the number of network hierarchy levels. The shift to IP networks also raises questions of whether interconnection frameworks need to be revised, such as a shift to use interconnect frameworks which have been successful in developing Internet markets. With the range of technologies making demand on spectrum, such as HDTV, mobile TV or 3G services, growing rapidly, the need to have a framework for allocation and management that can flexibly reassign unused and underused spectrum is becoming increasingly important. Here we briefly discuss this issue.

4.3 Interconnection Frameworks

Interconnection is essential in a competitive communications environment since it provides the means to allow the customer of any one communication service provider to connect with the customer of any other communications provider, and any service provider to connect, and provide service, to a customer irrespective of their network carrier. The transition to IP-based next generation networks is likely to raise questions as to how interconnection should be take place, given the significant differences in interconnection practices between the PSTN and IP networks and the fact that there will be interconnection between diverse networks including cable networks and the development of new services such as fixed-mobile converged services.

In the PSTN environment traditionally service providers adhere to wholesale payment arrangements known as *calling party's network pays* (CPNP), where the network of the party that places (originates) a phone call makes a wholesale payment to the network of the party that receives (terminates) the call. In contrast, Internet interconnection has been based on *peering*, *paid peering*, and *IP-transit*. With peering, two *Internet service providers* (ISPs) agree to exchange traffic solely among their respective customers, sometimes without payment; with transit, one ISP agrees to carry the traffic of a customer (possibly also an ISP) to third parties, generally for a fee. These arrangements based on commercial agreements result in an interconnected Internet, and have generally not been subject to regulatory obligations. The model that applies is therefore determined, in practice, by the type of interface used to exchange the traffic. The question is therefore on which model interconnection in a converged NGN environment should be based.

Depending on the strategy of companies, there will be a transition phase during which it is likely that both sets of practices will coexist as the proportion of IP traffic increases, and that of circuit-switched traffic decreases.

This may imply, as well, a transition in interconnection procedures. In many countries regulators use *long-run incremental cost* (LRIC) models to determine interconnection costs. There is a need for regulators to assess how the two sets of interconnection arrangements operate in their current milieus to evaluate whether these should be maintained in an NGN environment. The market for exchange of IP traffic, as regards the Internet, has worked well, producing efficient arrangements and lower prices, and allowing for entities of different sizes to exchange traffic [34].

In terms of physical facilities supporting traditional fixed and mobile switched interconnection, the migration towards NGN changes the network topology which potentially involves several structural changes, such as a reorganization of core network nodes and changes in the number of network hierarchy levels [35]. As an example, in Germany, Deutsche Telekom has 74 nodes for its IO network compared to 475 nodes for the PSTN [36]. This may lead to a geographical re-arrangement of points of interconnection, and to the reduction in the number of points, especially at the local level. At the same time it can result in new entrants being subject to stranded investment requiring them to invest in new infrastructure in order to reach new points of interconnection. There are different fiber network topologies in a NGN access environment which also may need to be taken into account since the requirements and points of interconnection may differ [37].

The separation of networks functional planes should allow for the creation of a horizontal platform for the provision of services, separated from the transport layer. For this separation to be effective, interconnection should be possible at all functional levels. However, there is the risk that operators do not consider horizontal separation appropriate, as it is more difficult to guarantee a certain level of quality of service in interconnected networks, or simply because it is not in their best interest. Most incumbent operators still see NGN as a simple continuation of vertically integrated transport and services, as in the case of legacy networks [38].

4.3.1 Numbering, Naming and Addressing

Telephone numbers, domain names, IP addresses, and other addresses are crucial resources for communication and access to the market. They provide operators and service providers with the necessary data for locating and identifying customers and network points in order to deliver their services. For end users they provide a presence in the world of communication and a means to communicate with others. For the PSTN, the public switched telephone network, the telephone numbering system [39], is the core mechanism to address end users. Practically all wire line and wireless networks operators base their interconnection, interoperability and service provisioning on the telephone system. With NGN, the existing numbering system is expected to continue, at least in the short to medium term, as the dominant scheme within voice communication to identify and connect subscribers.

Nevertheless, the same developments that characterize the merging communications landscape, such as the migration to IP, are affecting addressing as well, which raises risks in that access for users to competing service providers and/or services of their choice might not be achieved if the resolution between both addressing systems used (telephone numbers in PSTN, and IP addresses, domain names and *uniform resource identifiers* (URIs) in Internet) is not properly addressed with global standardization [40].

The IPv4 addressing scheme [41] as used in the Internet has been universally embraced by NGN networks as the core new addressing scheme, in combination with the overarching TCP/IP protocol suite [42]. IP addresses are used "under the hood" within networks and determinate unique network points; using an IP address will always lead to the exact location of that network point. On top of IP addressing there are translation mechanisms, such as the DNS (Domain Name System) that map or add other identifiers to an IP address. These identifiers, such as domain names, e-mail addresses and SIP addresses [43], are more comparable to telephone numbers, as they are used at the edges of networks, in the higher layer where services and applications take place in interaction with users.

With the expansion of the public Internet, the use of domain names and e-mail addresses for end users has become common practice worldwide, comparable to the expansion and acceptance of the telephone numbering system. Increasingly the underlying general format used in IP networks is the URI, the Uniform Resource Identifier. The URI is evolving into the main intranetwork identifier and basically defines an "identity-service" combination in a format like scheme:user@host or scheme:identifier@domain.tld. The URI format is versatile and, next to the well known URI for e-mail (mail to:user@domain.tld); the URI for SIP (sip:user@host) is becoming a main identifier to address VoIP subscribers according to the SIP protocol. These types of identifiers are all IP-based and can eventually be traced back to an IP address.

In parallel, other more closed identifier schemes have been introduced, mainly with the emergence of web-based VoIP and instant messaging (IM). Internet-focused companies such as eBay (Skype), Microsoft, Yahoo, Google and AOL have added voice, IM (instant messaging) and video capabilities to their software, serving large communities. They route mostly on the basis of "end to end point" communication, having the advantage that traffic does not need to be routed through the PSTN's traditional switches, or via SIP gateways as used within VoIP. These highly competitive providers on the voice market manage their subscribers' identities with proprietary schemes [44] and employ telephone numbering only when interoperability is needed with subscribers outside their community (Skype-in).

Although implemented on a provider by provider basis, IP-based schemes follow a standardized format and can be in principle supported across other networks. Interoperability is feasible if there is agreement between providers. The absence of interoperability is sometimes seen as a deliberate customer "lock in", as concluded by some parties on the basis that, e.g. Skype, will not map their end users to URIs, and the introduction of IP telephones that cannot be used for anything other than the application provided by the IP telephony provider.

Telephone numbers by which PSTN subscribers are identified may eventually evolve into alternative names and addresses, but generally many new services, such as web-based IM and VoIP services, are used "on top" of the regular voice subscription and this does not lead to the substitution of telephone numbers The emergence of new addresses, however, does lead to increasing divergence, as users are collecting more numbers and identifiers in different schemes, but there are no real indications that this divergence is posing problems on the end-user side; end-user equipment is becoming more intelligent and capable of handling multiple addresses and managing contact details.

The divergence however, does pose a challenge for providers. Telephone numbers in their standard format are not supported in the core NGN networks based on IP, where generally the URI format or other IP-based identifiers are used. Still, for users as well as for providers, being able to continue to use telephone numbers is considered crucial for the shift from the classic telephone service to VoIP and for the integration of new IP multimedia services. ENUM [45], a standard developed by the IETF [46] was conceived for this purpose; it offers a mechanism for transforming public telephone numbers into unique domain names. While solving the mapping problem, it introduced potential new applications, as a result of the insertion in the Domain Name System. ENUM comprises a set of standards and mechanisms for transforming public telephone numbers into unique domain names to be used in NGN, enabling providers and users to continue to use telephone numbers which is considered crucial for the shift from the existing public switched telecommunication environment to an Internet Protocol based environment and is thus becoming an essential building block for NGN embedded. Due to ENUM the lifespan of the existing telephone numbering scheme could be prolonged, subsequently maintaining the role of telephone numbers as key identifiers for telecommunication services. Eventually, however, regulators may need to introduce more flexibility in numbering plans by broadening the uses for existing number ranges, and considering portability of numbers between different services. At the same time access to ENUM data will become crucial to set up interconnection.

4.3.2 Universal Access and Next Generation Access

Convergence and the transition to next generation networks could, in the longer term, have an impact on the definition and scope of universal service obligations (USOs). At present USOs focus on the provision of voice services [47]. USOs generally refer to the requirement that a designated USO telecommunications operator provides a minimum set of services (which include voice telephone service) to all users, regardless of their geographical location within the national territory, at an affordable price, even though there may be significant differences in the cost of supply. Differently, the term "universal access" is used to refer to a situation where every person has a reasonable means of access to publicly available network facilities and services.

The communications market has been subject to significant changes both in terms of the means to provide voice services (mobile, VoIP) and the decreasing importance of voice services as a proportion of total telecommunications usage (e.g. because of e-mail, SMS, etc.). Many countries have stressed the economic and social importance of broadband access which in turn has led to considerations as to whether broadband access should be included as part of USOs. As the communications market evolves, particularly with regard to next generation networks, policy makers may need to review definitions of universal service to determine whether changes need to be made and, if so, what services and access would be required, and whether funding mechanisms should change.

The goal of universal service obligations generally is to promote the "availability, affordability and accessibility" [48] to telecommunications services. Definitions of universal service across most countries are relatively similar although there are differences in the mechanisms used to achieve these goals. Implicit in universal service goals in many countries is national tariff averaging aimed at assisting rural households (on the assumption that service costs are higher in those areas). In many countries part of USOs include, among other rates, special tariffs for those on low incomes.

Internet access is, to some extent, already included in universal service. For example, in the United States, the federal universal service schools and libraries program provides discounts for Internet access for schools and libraries throughout the nation, while the federal universal service rural health care program provides discounts to ensure comparability in Internet access rates paid by health care providers in rural areas and urban areas. In addition, the Federal Communications Commission (FCC) has initiated a universal service rural health care pilot program, which seeks to stimulate deployment of the broadband infrastructure necessary to support innovative tele-health and, in particular, telemedicine services to those areas of the United States where the need for those benefits is most acute. The European universal service directive (USD) [49] specifies that connections to the public telephone network at a fixed location should be capable of supporting speech, fax, and data communications at rates sufficient for "functional Internet access". The provision of functional Internet access has been interpreted by the Directive as encompassing simply the provision of a "narrowband connection" [50], and no minimum data rate is mandated in the directive. Overall, it seems that most EU countries opted for not requiring more than a 28 Kbit/s connection.

The definition of universal service is an evolving concept which may change over the years, to reflect advances in technologies and usages. For example, in the United States, universal service specifically is defined as "an evolving level of telecommunications service that the [FCC] shall establish periodically ... taking into account advances in telecommunications and information technologies and services" [51]. In the EU, to ensure that the changes in USO designations justify the important associated policy interventions, the Universal Service Directive established a number of criteria for modification. These usually include the popularity of the service, the diffusion of the technologies, and the likeliness that the unavailability of the service causes social exclusion. They also include considerations regarding "technological feasibility", the possibility to find "practical and efficient implementation mechanisms", and the balance between the cost of the measure and the benefits it will bring to society, always seeking to minimize market distortions [52].

4.3.3 NGN Lawful Interception

Subject to national legislation, all kinds of telecommunications may be subject to interception and/or data searches in relation to enquiries. *Lawful interception* (LI), also called "wiretapping", consists in the interception of communications by *law enforcement agencies* (LEAs) and intelligence services. The requests are directed to public telecommunication networks and services, in accordance with national legislation and on the basis of the authorization from competent authorities. With technological evolution it has become more difficult to intercept all communications of a targeted user. In the *public switched telephone network* (PSTN) environment interception was carried out by connecting to the line of the user at the local switch. With the advent of mobile phones, it became more difficult to implement lawful intercept since users could be at any location served by the operator and its roaming partners. The mobile signaling networks need to be monitored to detect the presence, identity and location of callers. On the technical side standards organizations have played a role in formulating standards which allow for lawful interception.

Convergence of networks and services, with users transmitting information through IP-based, mobile or fixed networks interchangeably, is exacerbating the problem of lawful intercept. To preserve the ability of law enforcement agencies to conduct electronic interception, network operators and application service providers, as well as manufacturers of telecommunications equipment, are required to modify and design their equipment, facilities, and services to ensure that they have the necessary capabilities to intercept. Governments extended the obligation to provide lawful interception from network operators to include also Internet service providers [53]. However, considering that often the reference is not anymore the connection, but the service used over the connection, questions arise as to whether the coverage of lawful intercept is adequate and whether this requires retention of data by, for example, Internet Service Providers. Communications using instant messaging or e-mail, as an example, do not necessarily have to be "home-based", but can use web-based mail where servers are located outside a country so that cross-border enforcement also becomes important. In this context, it is essential for law enforcement authorities to co-operate with network and service providers [54], as well as with application service providers, and to continue to work at the international level to build effective co-operation networks among different countries.

4.3.4 Fixed-Mobile Convergence

In the future network technology such as IMS (IP Multimedia Subsystem), should provide a standardized next generation architecture based on Internet Protocol (IP) for operators, and allow for the provision of mobile and fixed services using converged handsets embedding a radio interface such as cellular/Wi-Fi or cellular/Bluetooth dual-mode handsets. Currently, the main factor promoting FMC is the trend towards VoIP-enabled wireless telephony (VoWi-Fi), i.e. devices that use Wi-Fi to connect to a VoIP service such as Skype or roam between cellular and wireless LAN systems. Some of the VoWi-Fi operators are at present providing Wi-Fi based only services, but some are starting to offer FMC services by combining cellular services with VoWi-Fi. Challenges to mobile telecommunications operators are also coming from Wi-Fi hotspot operators, such as Bingo, allied with Skype. Some mobile operators are linking or considering linking their cellular networks with Wi-Fi hotspots and using VoWi-Fi to improve indoor coverage and offer low-cost calling in Wi-Fi locations. At present there various ways being used to provide FMC services, some of which are more technologically integrated than others. Dual-mode cellular/Wi-Fi handsets and using Wi-Fi modems in the home environment to access VoIP through ADSL connections can be found in some countries. There are less evolved forms of FMC using cellular/Wi-Fi dual-mode handsets that do not have a handover function or have a handover function but do not utilize a fixed voice or broadband network in the home. Services also exist linking both fixed and mobile networks which are not technologically converged, such as those offering a single voice mailbox over both fixed and mobile networks.

The deployment of NGN is expected to accelerate the offer of FMC services which are seamless to the user and use least cost routing. In turn, this may require that regulators review existing frameworks to ensure that they are not a disincentive to the development of new services, and that existing frameworks treat new services in a technologically neutral way. Numbering policies also have to accommodate FMC services and, if existing geographic numbers are used, then, in a calling party pays system, it may be necessary to devise ways to inform the call originator if different charges will be assessed based on the called party's location. It may also be important for regulators to develop adequate market tests given that the incumbents already have market power and often their mobile operators are also the market leaders; the development of FMC can augment this market power.

4.4 Broadcasting Convergence into IP-Based Networks

The digitalization of content, added to the shift towards IP-based networks, the diffusion of high-speed broadband access, and the availability of multimedia devices, allowed an increasing convergence of broadcasting and telecommunication sectors. The production and diffusion of audiovisual content does not seem to be limited to traditional broadcasters anymore. Telecommunication operators are providing content along with Internet access, newly emerging providers are offering access to content over IP, and traditional broadcasters are crossing over to other platforms, transmitting their programs also over IP networks.

Furthermore, the development of next generation mobile services – using 3G and 4G networks, or mobile broadcasting systems – enables the delivery of high quality audiovisual (AV) content to portable devices and mobile phones. Nowadays, convergence is a reality, with different types of content and communication services delivered through the same pipes and consumed over a variety of platforms and user devices. Convergence over multiple access platforms has not only affected the distribution market, but also created new forms of usage, providing consumers with greater choice and control over content. Multimedia, interactive audiovisual services are increasingly transforming users from passive watchers of TV programs to active players able to decide what they want to see, when and on which device. Video on Demand, Personal Video Recorder (PVR) services, peer to peer (P2P), or user-created video, therefore, herald an important change in the traditional broadcast model to exchange audiovisual content among large audiences. Media consumption, tastes and preferences may become more fragmented, the importance of social networks as a means to participate in content creation will probably continue to grow, and there will be an increasing demand for new types of content, able to fully capture the new capacity of the Internet for interactivity, non-linear consumption and participation [55].

The evolution of technology does not necessarily change many of the social and cultural broadcasting policy objectives, but technology may change the way that they are presently implemented and may allow for increased market liberalization than that which has been common in the sector while allowing the core policies to be maintained. The digitalization of transmission, for example, enables a more efficient use of spectrum than analog transmission, increasing significantly the number of terrestrial broadcasting channels which can be made available. When analog TV signals are switched off, a significant amount of spectrum bandwidth will be freed up, and will be available potentially for other applications, such as mobile television, high-definition television, mobile broadband networks and WiMAX networks [56]. Audiovisual content providers may include network operators, which are usually provide digital television and content over IP networks as part of their "triple play" bundles, or new service providers, such as Joost [57], using P2P technologies to stream content over the Internet, or YouTube, based *inter alia* on user created content. Broadcasters are also entering the IP market, launching new content platforms, such as Hulu-a NBC/NewsCorp venture (Table 4.3).

As the market for audiovisual services becomes more dynamic, content producers will be able to offer services directly to all new markets without intermediaries or gatekeepers. With content available on new platforms and networks, there should be lower entry barriers, and the sector could become more open and competitive over the next years. At the same time this will bring up the issue of the need for network neutral policy approaches, for both fixed and mobile networks, in order to avoid the creation of barriers to access for independent service providers [58].

In addition, existing government instruments to control broadcasting content – such as quotas for protection of language and culture, pluralism requirements, or must carry obligations – are challenged by the new multiplatform environment, and may need to be adjusted in order to continue to fulfill their goals.

Convergence not only leads to a larger and more competitive market, but also a more international market. A globally structured market – in terms of ownership, investment, and distribution and marketing strategies – offers an enormous potential to the media industries, but also poses new challenges to national regulation, which may not always be compatible across borders, therefore risking to be less effective, not enforceable, or – if excessively restrictive – to slow down growth of media players in an international content market.

4.4.1 Convergence in Content

While convergence may contribute to plurality and diversity, as it lowers market entry barriers, it creates new issues and challenges to existing policy. The telecommunication and broadcasting policy traditions may need to adjust in order to cope with the changing markets and to continue to achieve common policy objectives.

- *Scope of regulation*: Audiovisual content is increasingly distributed via a broad range of digital technologies that transmit to television, com-

Table 4.3 Examples of different ways to access content in a converged environment

	Provider	Content	Business model	Upload of user created content	Geographic restrictions
Managed IPNet	France Telecom	DTT + VoD (DSL)	Commercial + subscription channels	No	Yes
	BT Vision	DTT + VoD (DSL)	Commercial + subscription channels	No	Yes
New Internet Service/ Application Providers- IPTV Model	Joost	Streaming Independent/ Private content producers	Ad-supported	No	Yes- geographical blocking depending on content rights
	Babelgum	Streaming Independent/ Private content producers	Ad-supported	No	Yes- geographical blocking depending on content rights
Broadcast Operators	Hulu (Beta) NBC/ New Corp	Premium content from NBC/Fox + content from 15 other cable channels. Streaming from the main site or distribution partners (Yahoo, MSN, Comcast, AOL, MySpace)	Ad-supported, banners alongside the video, text along the bottom of the picture or clip	No	Yes: cannot access the service from outside the US
	BBC	On demand 7 day catch-up of BBC TV and radio programming	–	No	–
	YouTube	User created content, short professional video/trailers, promotional materials	Ad-supported targeted adverts, banners, etc.	Yes	No
	iTunes	Download of movies, music and podcasts, distribution agreements with content producers	Pay per download, free content is also available	Through podcasts	Yes, cannot download movies outside the US

puters, as well as mobile and portable devices, blurring boundaries between "video" and "broadcasting services" [59]. The scope of the definition of broadcasting services102 is relevant considering the detailed regulation which is usually imposed on broadcasters and usually aimed at addressing a number of social and economic interests, such as the need to maintain plurality and cultural diversity, develop national identity, and implement certain standards of decency. Policy makers need to determine whether and to what extent existing broadcast regulations should apply or be adapted to a wider range of content packagers and suppliers, and to what extent existing broadcast regulation may be reduced.

- *Ensuring effective competition*: Convergence is helping to intensify competition in broadcast markets by impacting on delivery networks and services. Convergence can help reduce access bottlenecks by allowing services to be delivered on a number of different platforms, and by creating market entry opportunities by new providers stimulates innovative services. The entry into the audio-visual market by new players, such as telecommunication network operators and larger Internet-based companies, can reduce market power in broadcasting. However, access to content is important for new entrants so that if larger companies or joint ventures (horizontal integration) control media rights for the most interesting premium content, it may be difficult for new entrants to provide competitive offers [60]. In addition, the development of some of the new technologies and services depends on the spectrum which is made available. With the shift to digital television more spectrums will be freed up and will be available for other services. The allocation of the so-called "digital dividend" can therefore have an impact on the development of new services in the content market [61]. Currently it seems that the request for spectrum will be driven by mobile television, high-definition television and wireless services, such as WiMAX.

- *Spectrum allocation*: The switching off of analog TV signals and the shift to digital transmission will make a significant amount of spectrum bandwidth available (the so-called "digital dividend"), which could be used for the provision of enhanced TV services, more TV channels, or some advanced wireless communication services [62]. In particular, the availability of spectrum to develop new wireless networks could help new entrants to create alternative access infrastructures and deliver directly their services to users, competing with incumbent operators.

- *Public interest objectives for content*: The rationale for special regulation of broadcast content is changing along with digitization and increasing access to on-demand audio-visual services. There is more choice, and an increasing proportion of consumers can now control the time of consumption of content. It is important, in view of the changes in the supply of information and programming, to reconsider how public interest objectives can be achieved in the digitalized IP world.

- *Advertising*: Advertising quotas and time frames were developed to limit commercial communications in traditional linear, point to multipoint broadcast transmissions. In a more interactive environment, and with VOD and PVR providing some possibility to skip frames, the traditional advertising model has become less effective. A controlled liberalization of some rules for television advertising, such as product placement, interactive online selling and banners during certain programs, could help the development of new business models, allowing broadcasters to compete with innovative Internet-based video services.

- *Must carry regulations*: Some countries enforce a certain form of *must carry* regulation. Most of these regulations were formed when there was a scarcity in distribution networks. As a result of technological and market developments, there is less dependence on a single infrastructure, and more channels and platforms for distribution of content are now available. Must carry rules should therefore be limited to a reasonable number of channels, including especially public service channels. Instead of "must carry", consideration could be given to a framework whereby terrestrial broadcast channels should be subject to "must offer" requirements, i.e. certain broadcasters are obliged to offer their content to other platforms if they ask for it.

• *Mobile video content*: The limitations and the cost of offering television on 3G networks using multimedia broadcast multicast service (MBMS) have encouraged operators to try to obtain separate allocations of spectrum for mobile television using a number of technologies. In addition, the interest for the allocation of new spectrum bandwidth may also be a means of pre-empting competition from broadcasters offering mobile television, and push convergence from the network into the handset.

The mobile television market does not seem to have deployed its full potential yet and innovation has been lagging behind, with sometimes restrictive platforms adopted by the wireless carriers and phone manufacturers. The EC estimated that the market for mobile TV would reach Euro 20 billion by 2011. However, it seems that mobile operators still have difficulties in identifying the appropriate business model for the service. Currently, revenues for mobile TV mainly come from the subscriptions, as advertising is not expected to be significant because of the low usage.

In November 2007, Google announced the launch of a new mobile operating system called Android. Based on Linux, Android provides an open platform for developers to create their own applications for a wide range of mobile devices, and will be available for free to cell phone manufacturers. Mobile-tailored content with targeted advertisement could therefore be one of the future models for mobile television. Another model for mobile television could be Qualcomm's one-way, multicast video programming network, MediaFLO. The MediaFLO service is currently offered by one mobile operator in the United States in approximately 40 US markets.

4.4.2 Emerging Trends in Service Delivery

Customers are now driving the market. It is the changing life style of the users that is making the device vendors to come up with devices that gives the flexibility. The new devices (iPhone, Palm Pre, Win Mobile) are giving choices to the customers to drive the market. For example, iPhone is a disruptive technology which created its own single eco system making the users build applications for it. The requirement of having one device for any kind of activity like voice calls, SMS, MMS, browsing, gaming, contents while being on any network and requiring a single bill for the usage is now taken for granted. This forces the service providers to move to the next generation systems, which is the convergence of devices, networks, and businesses.

Next generation operational support systems (OSS)/business support systems (BSS) are being evolved to manage the converged IP networks. The service providers' mission as a surviving service provider in 2009 and beyond is to leverage assets to bring customer winning products to market faster and better than their competitors at a price point that will keep customers "sticky". Doing so means providing and maintaining ubiquitous connectivity and service delivery to a burgeoning line-up of consumer and business devices – laptops, TVs, mobile handsets, PDAs, MP3/personal entertainment units – each with an array of functions that grows with introduction of each new model. The providers must enable users to mix, match and bundle services – whether that includes voice, video, data, wireless, entertainment, hosting and messaging or premium/lifestyle contents – at home, at work or in transit, and be ready to receive orders however customers wish to place them.

Next-generation *operations support systems* (OSSs) and *billing support systems* (BSSs) hold the key to service providers' new Holy Grail, which is the ability to deliver "any service to any device over any network, anytime, anywhere". But it takes a *service delivery* platform (SDP) to open the door.

Figure 4.3 presents an integrated OSS/BSS system. It consists of various components such as: InAdaptors, OutAdaptors, Transformers, Web Services, queues and BPM tool. A brief description of the components is provided below:

- *Business process management (BPM) layer*: it is used to define automated business processes. BPM systems can interact with the external systems through its off-the-shelf or custom-developed connectors. There are various players offering BPM tool like IBM, BEA, Vitria and Jboss. Most of the BPM tools deploy the business processes on a J2EE application server as *enterprise java beans* (EJB) and the whole application

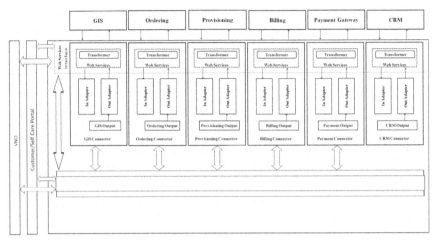

Figure 4.3 Integrated OSS/BSS architecture

is assembled into an EAR. So these business processes have access to all the J2EE server features like security transaction, JNDI, remote connectivity, etc.

- *Application queue*: the business processes defined in BPM can be invoked by sending a message in the application queue. The application queue is deployed as JMS queue on application server in order to provide asynchronous communication.
- *Connector*: it consists of InAdapter, OutAdpter, Web Service, Transformer and a connection-specific queue. Connector can be invoked either by placing the request in queue or by directly calling the web service. Connector is the only way through which the BPM layer can interact with the external system.
- *Connector Queue*: all the asynchronous requests that need to be handled by OutAdpter are placed in the Connector queue. Each connector has its own queue that is deployed as IMSqueue on the application server.
- *InAdaptor*: it is used to place a request in application queue that will invoke a workflow of BPM depending on the request type. For example, if account information has been updated in CRM system by CSR, it should be synchronized with the system. In order to accomplish the CRM INAdaptor will place an *updateaccount* request in the application server.

- *OutAdaptor*: it is used to process the requests that are placed in connector queue. In the CRM example, InAdaptor will place an *updateaccount* request in the application queue. Based on the message type, this request will go to the Billing queue, and then OutAdaptor will read the request, and call the web service to create an account in the billing system.
- *Transformer*: it is used to convert the object of a third party system into application-specific object and vice versa. It has only the transformation logic and no business logic. It should be able handle the following transformation: (i) Java object to Java object, (ii) Java to XML or vice versa, and (iii) XML to XML.
- *Web service interface*: the functionality of the external systems can be invoked through the web service interface. The Web Service interface will accept the application-specific parameters and transform them to external system-specific parameters and will then call the functionality of the external system. Similarly, the result will be transformed and returned.
- *Customer/Self care portal*: it consists of custom GUI that interacts with Integration Module.
- *VNO*: a virtual network operator (VNO) who wants to use the services provided by another network access provider (NAP) may directly interact with either exposed web service layer or the BPM layer.

4.5 Security in Converged NGN

The convergence of networks towards all-IP architecture provides operators with great opportunities to reduce their costs, and develop integrated services across fixed and mobile access increasing subscriber welfare. Network convergence needs to be complemented by convergence in the underlying security of policies, measures and practices so as to defend against various attacks [63]. As operators move from trials to wide commercial roll-out, questions regarding how to guarantee security across multiple networks are becoming more urgent.

IP-based next generation networks and the traditional circuit switched networks operate in different environments and are therefore exposed to different types of threats and attacks, both from within or outside the networks. With converged networks, operators are migrating from a stand-alone *closed* environment, such as the PSTN, to an open environment. The PSTN infrastructure is controlled by operators, and users have a lesser amount of information on its structure and functioning, as well as fewer possibilities to misuse the network. This situation, sometimes labeled as *security by obscurity*, [64] stands in contrast with the design of the IP infrastructure, based on open protocols, which were not originally designed for security implementation [65]. IP networks enable the provision of services – such as voice, data, and multimedia-provided by multiple access and service providers, and connected with a growing number of devices.

Security in a converging environment is not only a technical, but also an economic and social issue. On the economic side, networks are an integral part of the global information infrastructure, defined as an essential, indispensable facility for society, whose disruption would rapidly bring about a state of emergency or could have adverse societal effects in the longer term [66]. While governments and businesses are eager to adopt innovative services and applications; they require appropriate levels of assurance to protect their information and transactions. The social dimension of security is also important as convergence of networks and systems will expand opportunities for consumers to be connected anytime, anywhere. While the growing empowerment of users enables them to benefit more from ICTs, it also brings along with it increasing security and vulnerability risks for their transactions and personal information.

The borderless nature of IP networks means that security threats affecting the converged infrastructure can arise from anywhere. The main challenges across borders include the necessity to improve cooperation of law enforcement activities against security offences, with particular attention to consistency of cyber-crime legislation and regulations. In addition, international co-ordination and exchange of information is essential to create a global understanding of security risks and solutions linked to converged networks.

Although security is a priority in the future networks, it is also important to ensure an appropriate balance between civil liberties and security solutions – at the technical, policy or regulatory levels – in order to avoid excesses leading to violation of users' privacy, or illegitimately limiting individuals' rights to anonymity and freedom of expression [67]. It is also important to take into account the direct and indirect costs which may be incurred from securing networks. These costs also reduce the openness of networks and may impact on innovation.

International *standard development organizations* (SDOs) such as ITU, ETSI, ISO, IETF, 3GPP/3GPP2, are currently working to integrate security into the definition of NGN standards and protocols, in order to appropriately address security in the design phase of the new generation of networks. A set of specifications for IMS standards has been included in IMS Release 7, while TISPAN, in the preparation of its NGN Release 1, has been working on an equivalent set of specifications for broadband fixed access. TISPAN aligned its security approach with 3GPP where convergence was identified, adding TISPAN-tailored security specifications in areas where there are differences between fixed and mobile architecture. For example, pure wireline solutions do not have the same vulnerability as the mobile interface, which allows for the introduction of simplified security scenarios; on the other hand, fixed networks have to support inter-working with many sets of more or less secure protocol stacks, and with a wider variety of access technologies compared to mobile operators. In addition, user equipment vulnerability is more pronounced in fixed than in mobile networks, as users can modify their equipment without prior notice to the provider.

In a layered architecture, such as that of NGN, where services are separated from transport and access is enabled from multiple devices, security has to be considered at different points in the NGN architecture. In its NGN Release 1, ITU stressed the need to provide security of end-users communications across multiple-network administrative domains and identified three security layers: infrastructure security, service security and application security [68].

NGN solutions vendors also address the problem of security at different layers. These include access security, addressing direct or indirect connectivity of networks to user equipment (UE); intra-domain security, which is under the responsibility of the operator of the domain in question; and inter-domain security, i.e., security risks and threats associated with interconnection with untrusted networks [69]. In the latter case, security policies [70] from the originating network are usually enforced towards the destination network domain thanks to the utilization of *security gateways* (SEG) situated at the borders of different domains and communicating during interconnection.

Example: A specific example of possible security issues in an NGN environment can be provided by Voice over IP services. Voice is a critical service which in the past has benefited from separate PSTN and mobile networks, and had a certain degree of reliability. Shifting from PSTN to IP, the existing redundancy may be lost due to network convergence, and VoIP may inherit many of the problems already experienced by TCP/IP protocol data communications, such as attacks on confidentiality, integrity, availability and authenticity. Some of the current threats include the transmission of viruses and malware, eavesdropping, denial of service (DoS) attacks (Table 4.4). Al-

Table 4.4 Threats and risks in VoIP (Source: Trusted Information Sharing Network (TISN), Security of Voice over Internet Protocol: Advice for Chief Information Officers, September 2005. URL: `http://www.dcita.gov.au/communication_for_business/security/` `critical_infrastructure_security.html`)

Threat	Risk Issues
Eavesdropping through interception and/or duplication	Access can be gained through any access point to the voice network (Particularly if there are wireless access points in the same network that supports the VoIP service). Once access has been gained, network sniffer tools are commonly available to intercept IP-based traffic.
Loss, alteration or deletion of content	Exposure to programmed attack e.g., programmed substitution of Dual-Tone Multi-Frequency (DTMF) or Interactive Voice Response (IVR)
Caller ID/location may not be identified in an emergency	Complex numbering schemes, combined with incorrect PSTN access point routing, may provide wrong location information to emergency services. There is a greater risk of this happening when calls from remote offices are routed over a wide area network (WAN) before reaching the PSTN.
Lack of capacity/system management	Other network traffic can impact on VoIP traffic.
Denial of service attack	Swamping of network traffic resulting in no capacity to support voice. Can be targeted from within the enterprise or externally.
Viruses and other malware	Swamping of network traffic resulting in no capacity to support voice. Can be targeted from within the enterprise or externally. Viruses can also target specific VoIP protocols.
Power failure	VoIP is different from traditional telephony in that voice services are potentially vulnerable to a number of power failure points within the data network, e.g. local router and switches. In contrast, traditional telephony handsets are powered from one centralized point, usually with a backup battery bank.

though, operators are currently working on secure solutions for VoIP, service providers believe that it may be difficult to implement security while maintaining an appropriate level of Quality of Service (QoS), because of the extra processing and possible delay in communication it may cause [71].

An issue which may need to be specifically addressed in the context of NGN security is identity management, which in the NGN field has been technically described (at the working level) as the management by NGN providers of trusted attributes of an entity such as a subscriber, a device or a provider [72]. In a converged environment users would be able to use a single authentication mechanism (sign-in) on any access point on the NGN. The development and implementation of an authentication mechanism that allows a single and secure identification while protecting the privacy of the users however, is a great challenge [73]. In an environment with multiple providers, a common authentication process is difficult to achieve. However, a common authentication mechanism is also crucial in order to maintain a relationship between users, devices, and service and access providers. In addition, interoperable identity management is an issue that spans all layers from infrastructure to applications, and requires both technical and regulatory approaches harmonized at the international level [74].

4.6 IP Multimedia Subsystems (IMS)

Recently, web-based multimedia services have gained popularity and have proven themselves to be viable means of communications. This has inspired the telecommunication service providers and network operators to reinvent themselves to try and provide value added IP-centric services. There was a need for a system which would allow new services to be introduced rapidly with reduced capital expense (CAPEX) and operational expense (OPEX) through increased efficiency in network utilization. Various organizations and standardization agencies have been working together to establish such a system. Internet Protocol Multimedia Subsystem (IMS) is a result of these efforts. IMS is an application level system. It is being developed by 3GPP (3rd Generation Partnership Project) and 3GPP2 in collaboration with IETF (Internet Engineering Task Force), ITU-T (International Telecommunication Union-Telecommunication Standardization Sector), and ETSI (European Telecommunications Standard Institute), etc. Initially, the main aim of IMS was to bring together the internet and the cellular world, but it has extended to include traditional wireline telecommunication systems as well. It utilizes existing Internet protocols such as SIP (Session Initiation Protocol), AAA (Authentication, Authorization and Accounting protocol), COPS (Common Open Policy Service), etc., and modifies them to meet the stringent requirements of reliable, real-time communication systems. The advantages of IMS include easy service quality management, mobility management, service control and integration.

At present a lot of attention is being paid to provide bundled up services in the home environment. Service providers have been successful in providing traditional telephony, high speed Internet and cable services in a single package. But there is very little integration among these services. IMS can provide a way to integrate them as well as extend the possibility of various other services to be added to allow increased automation in the home environment.

This section extends the concept of IMS to provide convergence and facilitate inter-working of the various bundled services available in the home environment, which may include but is not limited to communications (wired and wireless), entertainment, security etc. In this section, a converged home environment is presented which has a number of elements providing a variety of communication and entertainment services. The proposed network would allow effective inter-working of these elements, based on IMS architecture. The objective is to depict the possible advantages of using IMS to provide convergence, automation and integration at the residential level.

4.6.1 IMS Architecture

The IMS comprises a core network (CN) which is a collection of signaling and bearer related network elements. These CN elements operate collectively to provide multimedia services to the end user. The IP multimedia services are based on the IETF defined standards for session control and bearer control. The IMS terminal connects to CN via an IP-Connectivity Access Network (IP-CAN), which functions merely as a means to transport IP data. This allows IMS to achieve *access independence* as defined in 3GPP (TS 22.228 V7.2.0 [75]. The access independence refers to the ability for the subscribers to access their IP multimedia services over any access network capable of providing IP-connectivity, e.g. via:

- 3GPP (UTRAN, GERAN).
- Non-3GPP access with specified interworking (e.g. WLAN with 3GPP interworking).
- Other non-3GPP accesses that are not within the current scope of 3GPP (e.g. xDSL, PSTN, satellite, WLAN without 3GPP interworking).

To understand the functionalities of IMS the following definitions will be useful [75]:

- *IP Multimedia CN subsystem*: comprises all CN elements for the provision of IP multimedia applications over IP multimedia sessions.
- *IP Multimedia application*: an application that handles one or more media simultaneously such as speech, audio, video and data (e.g. chat, text, shared whiteboard) in a synchronized way from the user's point of view. A multimedia application may involve multiple parties, multiple connec-

tions, and addition or deletion of resources within a single IP multimedia session. A user may invoke concurrent IP multimedia applications in an IP session.

- *IP Multimedia service*: an IP multimedia service is the user experience provided by one or more IP multimedia applications.
- *IP Multimedia session*: an IP multimedia session is a set of multimedia senders and receivers and the data streams flowing from senders to receivers. IP multimedia sessions are supported by the IP multimedia CN subsystem and are enabled by IP connectivity bearers (e.g. GPRS as a bearer). A user may invoke concurrent IP multimedia sessions.

The high-level requirements of IMS are as follows [76]:

- *Support to establish IP Multimedia sessions*: IMS can provide the users with a variety of services, but the most basic and most important service is the audio and video communication. The IP multimedia session is designed to support one or more multimedia applications. The IMS system also ensures that there is no compromise or reduction in privacy, security, or authentication as compared with traditional systems.
- *Support for QoS negotiation and assurance*: The IMS provides support for QoS negotiation for IP multimedia sessions, both at the time of establishment and during the session by the user and the operator. It also ensures that end-to-end QoS for voice is at least as good as that achieved by the circuit-switched wireless systems.
- *Support of interworking with the Internet and CS domain*: Support for interworking with the Internet domain is an essential requirement. The IMS users will be able to access information, services and applications available through the Internet. The IMS users will have the ability to establish IP multimedia session with non-IMS users from the Internet and the existing circuit-switched systems including PSTN and cellular networks.
- *Support for roaming*: IMS will allow users to roam between different service providers' networks. There are established procedures to transfer signaling, authentication, accounting, and other service-related information between different IMS operators in a standardized and secure fashion.
- *Support for service delivery control by the operator*: The system requires strict control in terms of service delivery options. The operator will have control over all the services being offered to the users. General control policies enable the operators to monitor and control the bandwidth re-

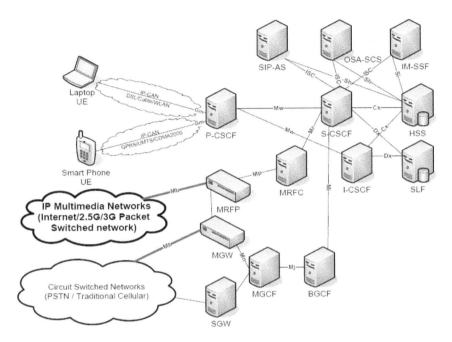

Figure 4.4 IMS architecture general overview

quirements in the network. By defining individual policies, the operators will have the ability to control sessions based on authorization.

- *Support for non-standardized rapid service creation*: IMS services need not be standardized. This allows the service providers or application developers to economically and rapidly develop and deploy services which would work equally well in different networks.

IMS CN Subsystem Architecture: Nodes and Their Functions

Before describing the IMS architecture, it is important to remember that IMS does not standardize a network element, but the functionality provided by the element. The manufacturer is free to decide about the physical design of the functional unit; two or more may be combined if deemed necessary. On the same lines, IMS does not standardize services but the service enablers.

The core network elements are shown in Figure 4.4. These nodes communicate with each other using specific protocols; each of these interfaces are identified using a reference point label as shown in Figure 4.4. A detailed list of all the interfaces and their operation is available in 3GPP TS 23.002 [77].

Another important characteristic of the IMS architecture is that it exclusively uses IPv6; it requires network elements such as NAT-PT (Network Address Translation – Protocol Translation) and IMSALG (IMS Application Level Gateway) to interoperate with the traditional Internet (which mostly uses IPv4). The description of the nodes and their role in the IMS environment is briefly presented below.

IMS Nodes: Databases

IMS utilizes one or more databases. The Home Subscriber Server (HSS) is used to store all the user related information, which is required to establish and handle multimedia sessions. The user information may include items such as the user profile (including the services the user subscribes to), location information, security information, the allotted S-CSCF address, etc. all the information is stored in a standard format and decisions are made by the HSS about the user sessions based on these items or user information. There could be more than one HSS, if the number of users is too high or for redundancy. In this case a Subscriber Location Function (SLF) is used to locate the HSS where the user information is stored. The SLF is a very simple data base which maps the user's address with an HSS, where all the user information is stored.

Both the HSS and SLF use DIAMETER protocol defined in RFC 3588 [78]. The DIAMETER protocol is the base protocol, there are specific applications developed for IMS to make the necessary decisions in the matters of authentication, authorization and accounting for a particular user.

Serving Call Session Control Function (S-CSCF)

The S-CSCF is a SIP based server and is one of the three types of *call session control functions* (CSCF). The basic function of a CSCF is to process the SIP (Session Initiation Protocol) signaling in the IMS [79]. SIP provides all the functionality to establish and manage multimedia sessions over IP networks. The S-CSCF is considered as the central node in the signaling plane. It is basically a SIP server but performs session control as well. It also maintains a session state as required by the network operator. Within a network there could be a number of S-CSCFs, with different functionality and used for different purposes. The main functions of the S-CSCF as defined in 3GPP TS 23.228 [80] are mentioned below:

1. *Registration related operations*

  - It may perform the task of the Registrar as defined in RFC 3261 [79]. It accepts the registration requests from the users, verifies the request by downloading the authentication vectors from the HSS. DIAMETER protocol is used for this purpose over the Cx interface.
  - It downloads the user profile from the HSS, which contains the service profile or information about any application servers which need to be included in the SIP procedures.
  - The S-CSCF makes the information available to the location servers, thus linking a particular user to an S-CSCF for the duration of the registration.

2. *Session related and session un-related flows*

  - It controls the sessions for the registered users and might deny establishment of different sessions (IMS communication) on the basis of various conditions or clauses that bar such an activity for that particular user.
  - The S-CSCF may behave as a proxy server as defined in RFC 3261 [79] or subsequent versions of the protocol. It may accept and service requests locally or forward them to the relevant node after translation and filtering the request.
  - The S-CSCF has the ability to behave as a User Agent as defined in RFC 3261 [79]; it can terminate and generate SIP transactions independently.
  - It can interact with different service platforms or application server over the ISC (IP Multimedia Subsystem Service Control) interface. This interface allows for the coordination and support of various services provided by the application servers.
  - The S-CSCF provides the endpoints with different service related information such as notification, location of additional media resources, billing notification, etc.
  - The functionality can be classified into two types; services provided for the originating end point and for the destination end point. First we look at the services provided for the originating end point which may include an originating user/User Equipment or an Application Server (AS).
    - The S-CSCF obtains the address of the Interrogating CSCF (I-CSCF) of the destination user from the destination name. This

is done in the case of the destination user being the customer of a different network. The S-CSCF forwards the request to the corresponding I-CSCF.

- If the destination user belongs to the same network the S-CSCF forwards the SIP request or response to the allocated I-CSCF with in the network.
- The S-CSCF also forwards the SIP request/response to a SIP server that is not a part of the IMS (e.g., internet). This depends on the policies of the home network operator.
- The S-CSCF forwards the SIP request or response to a Break-out Gateway Control Function (BGCF) for routing calls or sessions to the PSTN of any other Circuit Switched Domain (e.g. traditional cellular providers).
- If the incoming request is from an Application Server (AS), the S-CSCF will verify the request coming from the AS is an originating request and proceed accordingly. The S-CSCF will process and proceed even if the user on whose behalf the AS is acting is not registered and reflect in the charging information that AS initiated the session on behalf of the user.

- Services for the destination endpoint (terminating user/UE).

  - The S-CSCF will forward the SIP request or response to a specific Proxy CSCF (P-CSCF) as a part of the SIP terminating procedure to a home user within a home network or for a roaming user in a visited network.
  - The S-CSCF will forward a SIP request or response to an I-CSCF as part of the SIP terminating procedure for a roaming user in a visited network, where the home network operator chooses to include the I-CSCF in the path.
  - The S-CSCF will modify the SIP request as per directions for the HSS and the service control interactions, for routing the incoming session to the CS domain. This allows the user to receive the incoming session via the CS domain. It also forwards the SIP request or response to a BGCF for call routing to the PSTN or a CS domain.
  - The SIP request might contain preferences for the characteristics of the destination endpoints, the S-CSCF performs preference and capability matching as specified in RFC 3312 [81].

3. *Charging and resource utilization management monitoring*

- Like all the other nodes of IMS, the S-CSCF is a part of the complicated charging or accounting procedure. It generates Charging Data Records (CDR) for this purpose.

The S-CSCF is always located in the home network, and there are usually a number of S-CSCFs in a network for the sake of scalability and redundancy. Each of them can serve a number of IMS terminals at the same time.

Proxy-Call Session Control Function (P-CSCF)
As shown in Figure 4.4, the P-CSCF is the first contact between the user and the IMS network in the signaling plane. All the signaling and control information passes through the P-CSCF before getting to the user. It acts as an outbound/inbound SIP proxy server. The discovery of the address of the PCSCF and its allotment to the user is performed during the process of IMS Registration and this does not change for the duration of the registration. The P-CSCF discovery process is described in section 5.1.1 of 3GPP TS 23.288 [80].

The various functions performed by the P-CSCF are mentioned below:

- The P-CSCF establishes a security association between itself and the IMS terminal. The security association requirements and procedures are provided in 3GPP TS 33.203 [82]. The IPSec security associations provide integrity protection. The P-CSCF also authenticates the user and asserts the identity with rest of the nodes in the network, to avoid redundant authentication requirements.
- The P-CSCF forwards the SIP register request from the UE to the appropriate I-CSCF determined from the home domain name provided by the user. This allows for the successful IMS registration of the user/UE.
- The P-CSCF forwards the SIP request or responses to and from the UE to the allotted SIP server, which could be an S-CSCF. The address of the S-CSCF would have been obtained by the P-CSCF as a result of the registration process.
- The P-CSCF performs SIP message compression/decompression for the purpose of reducing the size of the messages and thus reducing the transmission time and quicker session establishment.
- The P-CSCF may also include a Policy Decision Function (PDF). It performs the task of authorizing the bearer resources and performing QoS management over the media plane. The PDF may not be included

in the same physical unit. Details about this function of the P-CSCF provided in 3GPP TS 23.207 [83].

- Like the S-CSCF the P-CSCF also generates the CDR and forwards the information to the charging collection node.

An IMS network may have multiple P-CSCF for scalability and redundancy. The P-CSCF may be located in the home network or in a visited network.

Interrogating Call Session Control Function (I-CSCF)

The I-CSCF is SIP proxy server; it is placed at the edge of the administrative domain of an IMS network. It is a point of contact for a connection destined to a user who belongs to that network, or a roaming user currently located within the service area of that network operator. The address of the I-CSCF is listed in the DNS (Domain Name System) database and is made available when a SIP server follows the protocol for locating a SIP server for the next hop, the protocol is provided in RFC 3263 [84]. The functions performed by the I-CSCF are mentioned below [75]:

- During the registration process the I-CSCF assigns an S-CSCF for a particular user. The ICSCF communicates with the HSS and SLF just like the S-CSCF over the Cx and Dx reference points. It uses the DIA-METER protocol. The user information is received by the I-CSCF and depending on the requirements; it assigns an S-CSCF to the user if one is not already allocated.
- The I-CSCF routes a SIP request from another network to the S-CSCF after obtaining the address of the appropriate S-CSCF from the HSS.
- The I-CSCF may encrypt certain parts of the SIP message which may contain sensitive information about the home domain; this function-ality is optional and is called THIG (Topology Hiding Inter-network Gateway).
- Like other CSCFs, the I-CSCF also generates CDRs to be transmitted to the charging collection node.

Application Servers

An application server provides value added services and can be located in the home network or any third party network. It is essentially a SIP sever which hosts and executes various services. There are different modes of operation for an Application Server. It can operate in SIP proxy mode, SIP User Agent mode (terminating or originating), SIP Back-to-Back User Agent (B2BUA) mode, etc.

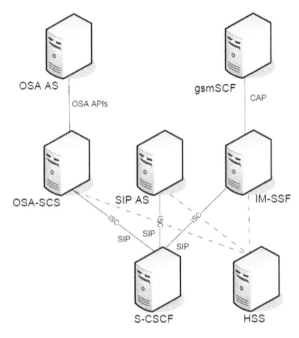

Figure 4.5 Types of application servers in IMS

The S-CSCF interfaces with the AS through the ISC (IP Multimedia Service Control) interface. The ISC interface is based on SIP [79]. Figure 4.5 shows three different types of Application Servers; they are described below.

- SIP AS: The SIP Application server is the native AS. It hosts and executes IP multimedia services based on SIP. All the new services that are going to be developed for the IMS architecture would be implemented using the SIP AS.
- OSA-SCS: The Open Source Access-Service Capability Server (OSA-SCS) provides an interface to the OSA framework applications. An AS located at a third party location will not be able to securely connect with the IMS network, whereas OSA has the capability to establish a secure connection with the IMS network. The OSA-SCS inherits all the abilities of OSA and is used to provide secure connectivity for a remotely located AS to the IMS network. The OSA-SCS acts as a regular AS and interfaces with S-CSCF via SIP on one end and as an OSA AS using OSA Application Programming Interface (API) on the other end. The OSA API is described in 3GPP TS 29.198 [85].

- IM-SSF: The IP Multimedia Service Switching Function is a specialized application server which allows for integration and reuse of the traditional applications developed for the GSM architecture. CAMEL (Customized Applications for Mobile network Enhanced Logic) was the name of the services that were developed to provide multimedia or enhanced services for GSM handsets. The IM-SSF acts as an application server on one side interfacing with the S-CSCF using SIP, and on the other side it acts as a Service Switching Function (SCF) interfacing with the gsmSCF with a protocol based on CAP (CAMEL Application Part). The CAP protocol is defined in 3GPP 29.278 [86].

The three AS mentioned above may perform different tasks but they behave exactly the same towards the IMS network. They all appear as SIP AS behaving in one of the earlier mentioned modes. The Application Servers present in the home network may optionally interface with the HSS. The SIP AS and the OSA-SCS interface with the HSS using DIAMETER protocol [78] and the interface is labeled "Sh". The IM-SSF interfaces with the HSS using protocol based on MAP (Mobile Application Part) defined in 3GPP TS 29.002 [87].

Breakout Gateway Control Function (BGCF)
As shown in Figure 4.4, the BGCF provides connectivity to the Circuit Switched domain through the MGCF (Media Gateway Control Function), SGW (Signaling Gateway) and the MGW (Media Gateway). These three nodes put together are referred to as the PSTN/CS Gateway. The BGCF is basically a SIP server which has the additional capability of routing and establishing sessions based on telephone numbers as user addresses. The BGCF is used exclusively for sessions initiated by an IMS user who needs to communicate with a user in the PSTN or PLMN (Public Land Mobile Network) domain, both of which are in the Circuit Switched domain. The main functions performed by the BGCF are as follows:

- It receives a request from the S-CSCF to select the appropriate PSTN/CS Domain break out point for a particular session.
- The BGCF selects the network in which the internetworking with the PSTN/CS Domain is to occur. If the interworking with PSTN/CS domain is to occur in the same domain, it selects the appropriate MGCF and forwards the SIP signaling to it.
- If the interworking with the PSTN/CS domain is to be done via a different network, the BGCF forwards the SIP information to the BGCF of

that network. If network hiding is required, the BGCF will forward the SIP signaling through the I-CSCF to the other BGCF.

- The BGCF also generates CDRs to forward to the charging collecting node.

Public Switched Telephone Network/Circuit Switched (PSTN/CS) Gateway
The PSTN/CS gateway comprises of three components as mentioned earlier. The functions performed by each of them are given below.

- *MGCF (Media Gateway Control Function)* interfaces with the BGCF and receives the SIP signaling. Its function is to convert the SIP signaling to either ISUP (Signaling System 7) defined in ITU-T Recommendation Q.761 [88] over IP or BICC (Bearer Independent Call Control) defined in ITU-T Recommendation Q.1901 [89] over IP. The converted signaling is forwarded to the Signaling Gateway (SGW). The MGCF also controls the resources in the Media Gateway (MGW). The MGCF and the MGW communicate with the help of the H.248 [90] protocol, specified in the ITU-T Recommendation H.248.

- *SGW (Signaling Gateway)* provides the signaling interface with the circuit switched domain. Its main function is to perform lower level protocol conversion. It converts MTP (Message Transfer Part) defined in ITU-T Recommendation Q.701 [91] into SCTP (Stream Control Transmission Protocol) defined in RFC 2960 [92] over IP. So the signaling format ISUP or BICC over MTP is transformed into ISUP or BICC over SCTP/IP.

- *MGW (Media Gateway)* connects the media plane of the PSTN or any other CS environment with the media plane of IMS. The MGW transcodes the IMS data transported over RTP (Real Time Protocol) defined in RFC 3550 [93] into PCM (Pulse Code Modulation) used in the PSTN environment. Also the MGW performs transcoding in situations where the IMS terminal does not support the codec being used by the CS side.

Media Resource Functions (MRFs)
The Media Resource Function (MRF) handles all the media transportation and processing requirements. It is divided into two functional components as shown in Figure 4.4, the Media Resource Function Controller (MRFC) and the Media Resource Function Processor (MRFC). The MRFC interfaces with the S-CSCF over the MRFC interface and uses SIP [57] for signaling purposes. The tasks performed by the MRFC are as below [80]:

- It controls the media stream resources in the MRFP.
- The MRFC interprets the information forwarded by the S-CSCF and the Application Servers and modifies the operation of the MRFP according to the directions.
- The MRFC generates CDRs like the other nodes in IMS to be forwarded to the charging collecting node.

The MRFP is controlled by the MRFC though the Mp interface, also called a reference point. The Mp reference point does not have a specific protocol specified for it yet and has an open architecture to allow extension work to be carried out. It completely supports the H.248 Standard [90]. The tasks performed by the MRFP are given below [61]:

- The MRFP controls the bearer plane on the Mb reference point.
- It provides the functionality of mixing various incoming media streams in case of a conference call.
- It acts as a source of media streams or plays streams as for multimedia announcements.
- The MRFC performs all other media processing functions such as transcoding, media analysis, etc.
- It also provides floor control or manages access rights in a conference environment.

4.6.2 IMS Protocols

The protocols used in the IMS environment are derived from the internet and wireless domain (GSM/GPRS). 3GPP decided to use the protocols being developed by the IETF and ITU-T for the IMS and thus was able to capitalize on their expertise in designing robust protocols. A brief description of the various protocols used in IMS is given below.

Session Control in IMS

The protocol used for session initiation and control in IMS over IP networks is the Session Initiation Protocol (SIP) specified by the IETF. SIP is a text based protocol unlike other session protocols such as BICC and H.323. This makes it easier to debug, extend, and build services on it. One of the main reasons for choosing SIP was the fact that it is based on many familiar and successful protocols such as SMTP (Simple Mail Transfer Protocol) and HTTP (Hypertext Transfer Protocol). Also SIP follows the familiar client-server model. SIP makes is very easy to develop new applications, which is one of the requirements of IMS.

Authentication, Authorization and Accounting (AAA) in IMS

Authentication Authorization and Accounting (AAA) operations play a very important role in any network, especially in the IMS environment. It is of great importance to have an efficient and highly reliable mechanism to perform the tasks of authenticating a user's identity, authorizing the user to access the appropriate resources and making sure the resources and services consumed are logged accurately and billed correctly. IMS uses the DIAMETER protocol to perform the AAA operations. It allows different nodes to access retrieve or modify user information from HSS or SLF. The DIAMETER protocol is an improvement over the older RADIUS protocol [94]. The DIAMETER protocol is defined in RFC 3588 [60]. The DIAMETER protocol is used over different interfaces such as Cx, Dx and Sh. It consists of a base protocol and is used to develop various DIAMETER applications. These applications are extended and customized for a particular purpose or an environment. The different interfaces may use different DIAMETER applications to perform the various AAA procedures.

Quality of Service in IMS

Generally there are two models to provide QoS on the packet switched IP domain specifically the Internet. They are the Integrated Service model and the Differentiated Service (DiffServ) model.

Integrated Service model is defined in RFC 1633 [83], it provides end to end QoS. The endpoints request a certain QoS and the network grants it. The protocol used by the Integrated Services architecture is RSVP (Resource reSerVation Protocol), it is specified in RFC 2205 [95] and has been updated by RFC 2750 [96] and RFC 3936 [97]. Integrated Service works well in small networks does not scale well as the routers have to store state information about every flow and perform lookup before routing any packet. The second model of QoS solves some of the problems faced while using the Integrated Service model. The DiffServ architecture is specified in RFC 2475 [98] and RFC 3260 [99]. The DiffServ servers need to maintain minimum state information about the flows and enables a quicker treatment for the packets flowing through them. In this architecture the router is aware of the treatment that needs to be given to each packet; the treatment is referred to as the Per Hop Behavior (PHB). Each PHB is identified by 8-bit codes called Differentiated Service Code Points (DSCP). The DSCP information is carried by the packets in their IP headers. In IPv4 it is placed in the "Type of Service" field and in IPv6 it is placed in the "Traffic Class" field.

IMS allows many different end-to-end QoS models. All the models are described in 3GPP TS 23.207 [83]. The terminals may use link layer resource reservations methods such as PDP context reservation, or directly use protocols such as DiffServ or RSVP. The IMS networks use DiffServ and may use RSVP.

Security in IMS

Security generally deals with integrity, confidentiality, and availability. There are various means to achieve the security requirements in the SIP environment. In IMS security can be divided in two different areas, Access security and Network security.

Assess security deals with authentication and authorization processes and establishment of the IPsec security authorization (architecture defined in RFC 2401 [100]); these procedures are performed during the REGISTER transaction. All the procedures for security access are provided in 3GPP TS 33.203 [101]. In the 3GPP networks the user identity is stored on a smart card inserted in the IMS terminal; this card is usually known as UICC (Universal Integrated Circuit Card).

Network security deals with protecting the traffic between two nodes. The nodes may or may not belong to the network. There may be different levels of requirements from the network security mechanisms in place. If two different security domains are involved, the traffic travels through two Security Gateways (SEG). In this case the traffic is protected using IPsec ESP (Encapsulated Security Payload), specified in RFC 2406 [102] and runs in tunnel mode. The security associations are established and maintained using IKE (Internet Key Exchange), specified in RFC 2409 [103]. All the network security requirements are mentioned in 3GPP TS 33.210 [104].

Policy Control in IMS

Policy control deals with media-level access control; the decisions made by the policy control mechanism authorize a user to use the media plane and assigns the QoS to be provided for that user session. The media-level policy is enforced by the routers present in the network, but these routers do not have the ability to make decisions about users as they do not have access to the user information stored in the HSS. The task of obtaining the user information and making these decisions is performed by a SIP server in this case. The SIP server informs the routers to allow of deny a certain user with the requested media resources.

The node which makes the decision, in this case the SIP Server, is called the Policy Decision Point (PDP) and the router is called the Policy Enforcement Point (PEP). The protocol used between the PDP and PEP is called Common Open Policy Service (COPS) protocol, it is defined in RFC 2748 [105] and has been updated by RFC 4261 [106], which provides a higher level of security at the transport level. COPS generally supports two models for policy control, the outsourcing model and the configuration model (also called provisioning).

In the outsourcing model the PEP contacts the PDP for every decision, whereas in the configuration model the PEP stores the policy from the PDP locally and uses it to make decisions. IMS uses a combination of the two models, called COPS-PR. It is a mixture of the two models as it uses the same message format and the Policy Information Bases (PIB) as used by the provisioning model and the policy decision are transferred in real time like in the outsourcing model.

In IMS there are two types of limitations on the session that can be established. They are user-specific limitations and general network related policies. The user-specific limitations include restrictions on a particular user, in terms of resources that are allowed. An example would be an audio only subscription, so the user will not be allowed to establish video sessions. The general network policies would apply to all the users of that network. This might include restrictions on the codecs that can be used. The P-CSCF deals only with the enforcement of the general network policies, whereas the SC-SCF handles both user-specific policies and the network policies. Both these PDPs use the same mechanism to monitor the sessions. They access the SDP (Session Description Protocol) body to identify the type of session and media requested during SIP procedures.

4.7 Convergence Using IMS – A Case Study

In this section, we present an illustrative case study on convergence using IMS in a residential environment. First we describe the requirement of a converged residential environment.

Bundling up of communication services has been very successful of late. Companies have been providing Internet, voice telephony, and digital entertainment services to residential subscribers with the convenience of a single bill. These services include cellular telephony also. What lacks in this environment is interworking among various services.

A converged environment would allow for rich multimedia applications to be accessed by the users on any device and retain the user profile and other settings. The idea is to be able to communicate, establish multimedia sessions, and perform control and configuration operations on all the networked elements in the residential environment. The following are the typical requirements of a converged residential network:

- The system should allow multiple user profiles with different levels of control over services and the systems.
- It should also allow integration of wireline and wireless (cellular) voice communication, or should allow the users to make and receive phone calls on wireline-based digital/analog phone and the wireless cellular phone interchangeably as per personal preference.
- The system should allow for multimedia sessions to be established between the various terminals (audio/video database server, TiVo, etc.) in the network. The users should be able to access the content within the residential environment on any terminal or from outside the residential environment (restricted only by the capabilities of the device and the available bandwidth).
- The nodes in the residential environment may use either wired or wireless connectivity. The local area network should support secure wired and wireless access (Wi-Fi or WiMAX).
- The network must support application or service hosting and the application server should be remotely manageable. These applications may include but are not limited to residential security system (monitoring, configuring and authorization), web hosting services, residential power/gas monitoring and control, etc.
- The residential environment should be secure, easy to manage, and efficient, and should provide effective means of managing and configuring devices in the network.

Figure 4.6 shows such a converged architecture which shows a few of the many possible elements in the network. All these elements should be accessible to the user with sufficient rights, both locally and remotely. The users in the residential network subscribe to a number of services which might include cable television (digital or analog), Internet access, telephone connection, and one or more cellular phone subscriptions. As shown in Figure 4.6, we will assume these services are being provided by the Multimedia/Communication

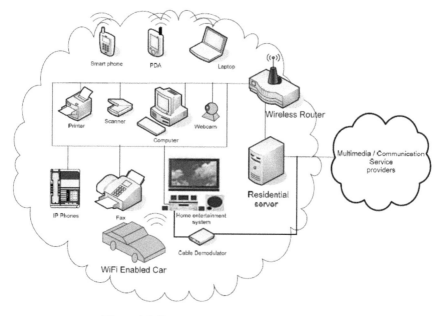

Figure 4.6 Converged home network environment

service providers. All the elements in the residential environment are networked as in a LAN environment. All there services are controlled by a so called residential server which performs the control operation and a wireless router which provides the connectivity. There could be other systems operational in the residence, such as a security system, an electronic power supply and control mechanism, which controls the lights, air conditioning and other energy related functions, etc. All these are linked and controlled by the residential server as well.

4.7.1 Scenarios in Converged Residential Network

We shall consider a family comprising four people, two parents and two children. Each of these four users might need different services and will have different levels of authority over the system. Let us consider a problem or a convergence requirement of this residential network and work on providing a solution for it. We assume that the father (John) has a cell phone (GSM/UMTS). He also subscribes to digital television services from the same company which provides him with a digital telephone and Internet access. This is a typical scenario at present for a customer of triple play services.

What John desires from this converged system is to have the following features:

- When at home, he should be able to make and receive calls (voice and maybe video) using his home digital phone or cell phone interchangeably. This means that once he gets home his cell phone calls should be automatically routed to his home phone to save minutes.
- The home phone is connected to a high bandwidth connection; it is an IP phone and should be allowed to be implemented over the Wi-Fi network. The cell phone used by John may be a smart phone which has Wi-Fi connectivity, so he can continue to use his handset to receive calls intended for both his home phone number and cell phone number but with enhanced bandwidth, improved display and reduced cost of access.
- Other members of the family may or may not have a cell phone account; if they do same should apply for them. Also the system has to be smart enough to identify the called party and not alert John on his phone if the call is for his children. This needs to work both ways as he does not need to his calls to be forwarded to other members of his family. The system should be configurable to manage this.
- He should be able to continue to access the internet from any other device, including a desktop, a laptop, a PDA/Smart phone, etc. Also he should be able to receive communication (voice call, email, voice mail, instant message, video call, etc.) addressed to his various accounts (email, home phone number, cell phone number, etc.) on any of the devices listed above. They have different abilities in terms of processing power, screen size, etc.
- The entertainment services being subscribed by the family may include digital TV, access to other online multimedia services, music, videos etc. The family must have access to the entertainment services from any network node capable of playing audio and video. This includes watching a particular TV channel on a PC, a laptop or a PDA.
- The family might like to share some data among themselves and their friends; this might be photos, video, audio or other data. They should be able to do that securely, from within the home and even outside home. John might use a database server which would be accessible to all authorized people.
- John would like to be able to add services or applications being run in the home network without changing the system much and should be able

to remotely configure and control those services. These might include a security system, a web server, etc.

The requirements mentioned above are quite advanced. However, the NGN system needs to be designed so that it is able to handle all these requirements and more. In the following section, we propose a mechanism of establishing such a system based on enhanced IMS architecture.

4.7.2 Achieving Converged Residential Network Using IMS

There are certain pre-requisites for establishing a session in the IMS environment in the residential network described in the previous section. We first present these re-requisites:

- *Establishing an IMS service contract*: This includes establishing a subscription with the IMS service provider. During this process, the service provider will provide the customer with the appropriate identities and the service profiles will be created depending on the user's requirements in terms of services, bandwidth for those services, and access to various other applications being provided by the service provider. After the service contract is established, the user profile will be stored in the HSS and will be used during various operations in IMS, including Authentication,, Authorization and Accounting purposes etc.
- *Obtaining and IP address*: Every IS terminal needs to get connected to the IMS core network. The connectivity is provided by the IP-CAN (IP-Connectivity Access Network). This could be any IP-based transport network such as GPRS (as in GSM/UMTS network), xDSL, Wireless LAN through Wi-Fi (IEEE 802.11) or WiMAX (IEEE 802.16) networks, etc. IMS uses only IPv6 address.
- *Discovery of P-CSCF*: After the IMS terminal obtains IPv6 address, the next step is to locate a P-CSCF. This procedure includes the discovery of the IP address of the P-CSCF, which acts as an inbound/outbound SIP proxy server and will interact with the IMS core network.
- *IMS registration*: The registration process in IMS is based on the SIP registration process, where a public user identity is bound to a SIP URI (Uniform Resource Identifier). This SIP URI contains the IPv6 address or the host name of the terminal where the user is reachable. This is done using the SIP REGISTER request defined in RFC 3261 [79]. The registration process in IMS needs to accomplish the following tasks:
    - Bind a Public User Identity to a contact address.

- Authentication of the user by the home network.
- Authentication of the network by the user.
- The SIP registration is authorized by the home network and allows for the usage of the IMS resources subscribed by the user.
- Verification of a roaming agreement between the home network and the visited network is the P-CSCF is located outside the home network, thus authorizing the usage of resources.
- The home network informing the user about the various other identities that have been allocated to the user (implicitly registered user identities).
- Negotiation of the security mechanisms between the IMS terminal and the P-CSCF for subsequent signalling and other such security association to protect the integrity of the messages.
- Uploading of the compression algorithms between the IMS terminal and the P-CSCF.

Assumptions:

- The users in the residential network are IMS subscribers and are assigned Public and Private User Identities by the IMS provider. The IMS provider provides the users with a UICC (Universal Integrated Circuit Card), which may contain an ISIM (IMS SIM) application, or an UMTS (UMTS SIM) application or both. The information stored on this card includes among other things are the Private User Identity, and one or more Public User Identities.
- The IP Connectivity Access Network (IP-CAN) used by the residential network is operated by the IMS provider.
- The residential network is an IP-based Ethernet environment, and uses IPv6 protocol.
- All the networked devices have IPv6 interfaces (NIC) and can obtain a globally unique IPv6 address.
- No traditional PSTN terminals are being used and all terminals are IP-based. However, the IMS terminals and users are fully capable of calling PSTN or other CS customers and vice versa.
- The entertainment services are digital and use configurable IP-based set-top devices.

IMS Implementation of the Residential Network

Figure 4.7 depicts the proposed architecture for the residential network. Since the IMS service provider also provides the IP-CAN, all the nodes in the

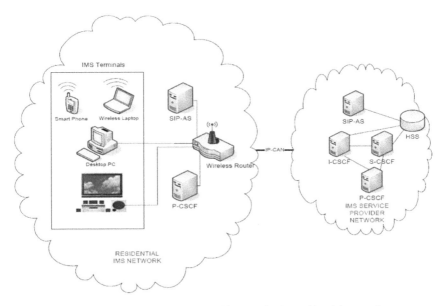

Figure 4.7 The proposed IMS Architecture in the residential network

residential environment can be considered to be a part of the Home (IMS provider's) network. The location of the P-CSCF and the SIP Application Server (SIP AS) is unorthodox in the network shown in Figure 4.7. However, they are technically still in the home network of the IMS provider. The architecture allows for a high level of flexibility in terms of the IMS service provider. The service provider can be changed without any major changes in the network, except for some reconfiguration of the nodes.

The IMS agreement is established to allow the user (in this example John) to register any of the public user identities at any of the available IMS terminals. The abilities of the terminal and the subscription of John regulate the type of sessions that can be established. The IMS terminals obtain an IPv6 address from the IP-CAN provider using the required DHCPv6 and DNS procedures.

The P-CSCF allotment is also regulated by the IMS service provider. In this case, we assume that the service provider has located an outbound SIP proxy (P-CSCF) at the customer's premises and this PCSCF is allocated to any sessions established by the user from the residential network. This may be done by configuring the DHCPv6 for a certain set of addresses or terminal names.

In the registration process, the user information stored in the USIM or ISIM (present in the UICC-Universal Integrated Circuit Card) would need to be accessed and transmitted in the right format from the IMS terminal. To accomplish this task, the IMS terminal being used would need to be connected in some way with the smart card allocated to the user. This smart card would probably be present in the wireless handheld device (probably a smart phone) being used by the user, and the user might perform this task by docking the smart phone using a cable or running a secure wireless application which transmits the relevant information from the smart card to the IMS terminal used. Once the authentication and authorization is complete and the S-CSCF is allocated to the user, it goes through the process of evaluating the initial filter criterion, which may result in invocation of one or more Application Servers.

Suppose that we need to provide convergence between the wireless cell phone used by John and the residential digital phone present in the residential network. We assume that there are two service profiles being used by John. Our task at hand when John is registered under the *home service profile* is to handle all incoming calls (voice, video or multimedia) to John's *general service profile* as per John's directive. He may wish for calls from certain people to be forwarded to his Home contact address (his residential number) and the rest of them might be forwarded to some sort of automated response and messaging system (answering machine); there are various other possibilities or ways in which John wishes to be reached while he is registered using his home profile.

This is achieved using the *initial filter criterion* present in John's user profile. Let us assume that John maintains a list of people (with known contacts) from whom he wishes to establish incoming communication no matter where he is registered (home or outside). This list is stored in one of the application servers, say *AS1*. Another Application Server, say *AS2*, receives input from AS1 regarding the session and acts as a SIP proxy and directs the session to the appropriate location. Both these Application Servers can be configured by the user over the "Ut" interface–the interface is between the User Equipment and the Application Server and is used exclusively for the purpose of configuring access related information and not for live traffic. The security functions for this interface are defined in 3GPP TS 33.222 [107].

When John registers at home, the S-CSCF invokes the *AS1* which has been configured by John as per his wishes and regulates his presence at home. This information is used as an input to *AS2* which receives the contact address of John and *AS1*. *AS2* will also have access to the list of people who would be permitted to contact John at home.

4.8 Convergence Standardization Organizations

The next generation networks (NGN) or advanced networks and systems are being developed in different parts of the world and by various agencies and organizations. Most of them are working in tandem to develop a uniform standardized system. The purpose of this section is to clearly identify the reasons for these efforts and their objectives. We will examine the requirements of the NGN systems as defined by ITU and will look at the various organizations working together to obtain a standardized solution for the same. As mentioned earlier, there are a number of agencies working towards the goal of advanced future networks. A major contributor in this direction is 3GPP and 3GPP2, which have introduced IMS. In the USA, ATIS (Alliance for Telecommunication Industry Solutions) is leading the way in developing NGN systems based on extended IMS. We will look into their operations in some depth.

The objective of ATIS, and more specifically ATIS Next Generation Network-Focus Group (NGNFG) is to design NGN which has the following fundamental aspects:

- Packet-based transfer.
- Separation of control functions among bearer capabilities, call/session, and application/service.
- Decoupling of service provision from network, and provision of open interfaces.
- Support for a wide range of services, applications and mechanisms based on service building blocks (including real time/streaming/non-real time services and multimedia).
- Broadband capabilities with end-to-end QoS and transparency.
- Interworking with legacy networks via open interfaces.
- Generalized mobility.
- Unrestricted access by users to different service providers.
- A variety of identification schemes which can be resolved to IP addresses for the purposes of routing in IP networks.
- Unified service characteristics for the same service as perceived by the user.
- Converged services between fixed/mobile.

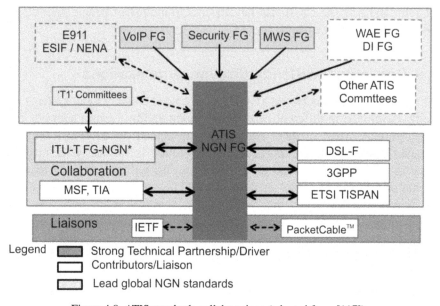

Figure 4.8 ATIS standards collaborations (adapted from [117])

- Independence of service-related functions from underlying transport technologies.
- Compliant with all regulatory requirements, for example, concerning emergency communications and security/privacy, etc.

ATIS NGN-FG is driven by the business needs of the North American market; the aim is to produce as much as possible international NGN standards. For this reason, it works in collaboration with a number of global standardization agencies.

Figure 4.8 shows the interaction of ATIS with other organizations and agencies involved in the standardization process of next generation converged networks.

The list of partners in Figure 4.8 is given below:

- ITU-T: SG13, including the Focus Group on NGN (FGNGN) 3rd
- Generation Partnership Projects (3GPP)
- European Telecommunication Standards Institute (ETSI) TISPAN
- Multiservice Switching Forum (MSF)
- DSL Forum (DSL-F)
- CableLabs

- Institute for Electrical and Electronics Engineer (IEEE)
- ATIS Technical Committees (e.g., PTSC, TMOC, NIPP) 3rd
- Generation Partnership Project #2 (3GPP2)
- Telecommunications Industry Association (TIA)
- Internet Engineering Task Force
- National Emergency Numbering Association (NENA)
- Emergency Services Interconnection Forum (ESIF)
- TTY Forum
- Industry Numbering Committee (INC)
- FCC Network Reliability & Interoperability Council (NRIC)
- Open Mobile Alliance (OMA)
- Metro Ethernet Forum (MEF)
- MPLS and Frame Relay Alliance

ITU-T study groups including SG-13 and its Focus Group on NGN (FGNGN) are responsible for the global standards for NGN telecommunication systems. The idea is to expedite the process of standards development by having regional Standard Development Organization (SDO) submit contributions to the IT-T. Such a workflow provides and establishes a global scope to the efforts of ITU-T and allows for a more harmonized approach.

TISPAN is the ETSI group responsible for all aspects of standardization for present and future converged networks, including the NGN and including service aspects, architectural aspects, protocol aspects, QoS studies, security related studies, and mobility aspects within fixed networks, using existing and emerging technologies. TISPAN has been developed architecture for NGN based on 3GPP IMS subsystem. This architecture is named extended IMS and is considered as the base for ATIS NGN work.

There are many other organizations and agencies which ATIS partners with to achieve its goals and objectives. Different agencies define standards for different mediums of communications, so it is essential for these organizations to interact and develop interoperable standards. For example, DSL Forum defines standards for data communication over the telephone lines, CableLabs develops standards for communication over the cable infrastructure and IEEE has been developing and defining standards for the data communication for the LAN/MAN, in both wired (IEEE 802.3) and wireless medium (IEEE 802.11-WiFi and 802.16-WiMAX).

4.8.1 Alliance for Telecommunications Industry Solutions (ATIS)

The ATIS NGN Framework documents Parts I and II provide a detailed set of requirements for the NGN systems. These documents examine every aspect of telecommunications and clearly define expectations from the NGN systems being developed. The requirements have been divided into six major groups:

- General requirements
- US regulatory requirements
- End user applications
- Network service enablers
- Underlying network/support capabilities
- Business model-driven requirements

The general requirements section [108] of the document contains guidelines about the basic functionality that would be required to exist for the NGN to fulfill the ITU-T requirements. It includes issues such as NGN network interconnection requirements, different types of interfaces between the *application service providers* (ASP) and the *next generation service providers* (NGSP), mechanisms to measure and predict QoS, guidelines for incremental replacement of legacy services, PSTN simulation, PSTN emulation, mobile network evolution, transparent end-to-end communication, synchronization and timing issues, etc.

The US regulatory requirements section of the document lays down all the possible regulatory requirements that might be enforced on the NGN systems. The regulatory requirements are a moving target and may evolve with the regulatory and legislative actions of the various government agencies at different points of time. These requirements include information and guidelines regarding various service-specific or general regulations such as *lawfully authorized electronic surveillance* (LAES), number portability regulations, number pooling, E 9-1-1 directives, *emergency telecoms service* (ETS), FCC rules and regulations, accounting, etc.

The end user applications section of the document provides guidelines regarding the user applications and about the way they would need to be approached in the new NGN network architecture. The NGN would require application to be supported on a common, converged architecture, so there would be need for changes or variations in the services and their inherent capabilities. This section lays down requirements for applications such as interactive voice, content and capabilities of video services, multimedia conferencing, content sharing, basic and advanced interactive gaming, sensor and control networking, mobility management, etc.

Some of the key *network service enablers* are defined in the next section of the document. Even though the NGN is not vertically integrated, there are still some network-based communication services that bring value to customer applications. These include QoS requirements, presence services, policy definitions and enforcement, media resource and media gateway functions, personal profiles unified interface, and service ubiquity. There are other important aspects which have been defined and described in this document such as roaming, location services, personal information management and access, digital rights management, and session management.

The underlying network/support capabilities are those that are not directly accessible by the applications. These are of monumental importance and need to be introduced in the core network and are very well-defined in this section of the document. These capabilities include *operations administration maintenance and provisioning* (OAMP), security provision, integrity requirements, confidentiality and privacy requirements, attack mitigation and prevention policies, accounting (ordering and billing), trust policies are requirements, etc.

The last segment of the requirements identifies and defines various business model-driven requirements. This section covers topics such as the operational expense (OPEX), implications for service providers, third part access implications, service delivery environment, and consolidated operations requirements.

The approach adopted by ATIS to develop NGN systems includes identifying a roadmap. The primary purpose of identifying the NGN roadmap is the creation of an infrastructure that enables flexible and efficient creation of NGN services. The focus of the roadmap is to identify the underlying network service enablers that will allow potential new services to be introduced. The roadmap is developed keeping in mind the heterogeneity in the industry in terms of technological (IP network capabilities) starting positions in moving towards NGN. The network service enablers required for developing new services suggested as the starting point of the NGN include the following:

- Unified user profile
- Security (various aspects as defined in NGN framework)
- Decoupling of services from access technologies
- Integrated management of all services, users and networks
- Presence

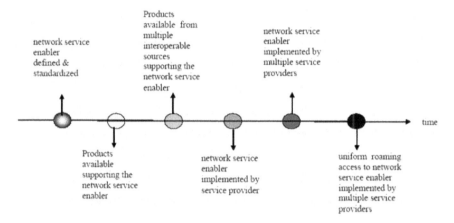

Figure 4.9 NGN capability deployment roadmap

- Scalable management and operations
- Quality of Service (QoS)
- Settlement (accounting)
- Digital Rights Management (DRM)
- Media Resource Functions (MRF), etc.

All these network service enablers have been assigned different priorities and have other functions dependent on them [109]. Figure 4.9 illustrates the flow of new network service enablers.

4.8.2 Internet Engineering Task Force (IETF)

As IETF is an important standardization body for Internet protocols and since it is playing an important role in evolution of IMS standards, we provide a very brief description about its activities. IETF is a loosely organized, self governed organization consisting of a wide variety of people with different technical backgrounds including network designers, vendors, etc. Their efforts are directed towards the development of the architecture, protocols, and the operations of the public Internet. Their mission statement is documented in RFC 3935 [110].

The IETF is organized into a number of working groups and the actual technical developmental work is done under one of the working groups. These working groups are organized into Area Directorates, and all these are managed by the Internet Engineering Steering Group (IESG). The IESG is responsible for the technical management of the IETF and decides the area the IETF should work on. The members of the IESG also review all the specifications that are produced.

The technical documents used within the group are called Internet Drafts. There can be two types of drafts: individual submission and working group items. The individual submissions are reviewed by the members and become group item if they are found worthy of investigation by the rest of the group. Work is done on the group item and is eventually submitted to the IESG, when the group is confident that it is ready for publication. The IESG provides feedback and approves the publication of a new RFC (Request for Comment). The internet drafts can be considered to be stable specifications only after they are RFCs.

4.8.3 Third Generation Partnership Project (3GPP)

3GPP and 3GPP2 are standardization bodies responsible for the development of the 3G cellular protocols and standards. Currently 3GPP is involved developing the Long Term Evolution (LTE) standards for next-generation cellular communication networks. 3GPP comprises a number of international *standard development organizations* (SDO) such as ARIB, ETSI, etc. 3GPP is organized into *technical specification groups* (TSG) and they are managed by a supervising organization called the *project co-ordination group* (PCG). The TSGs do not produce standards but they deliver *technical specifications* (TS) and *technical reports* (TR). Once these are approved by the TSGs, they are submitted to the organizational partners for the documents to go through their individual standardization processes.

3GPP2 has a very similar structure and operates in pretty much the same way. 3GPP2 version of IMS is called Multimedia Domain (MMD). 3GPP2 is organized into 4 TSGs. TSG-A deals with *access network interface*. TSG-C focuses on CDMA technology, TSG-S on service and systems aspects, and TSG-X on intersystem operations. 3GPP2 also delivers technical specification (TS) and technical reports (TR), but follows a different numbering pattern.

Almost all the protocols selected for IMS were originally from the Internet domain. However, there was an obvious need to modify the existing protocols to meet the requirements of IMS. For this purpose, IETF started jointly developing the protocols with 3GPP/3GPP2. A collaboration was established between these organizations and documented in RFC 3313 [111] (IETF-3GPP) and RFC 3131 [112] (IETF-3GPP2). The focus of collaboration in the Internet area was in the fields of IPv6 and DNS. One of the outcomes of these efforts was the development of RFC 3316 [55] by the IPv6 working group. This RFC provides guidelines for the implementation of IPv6 on cellular hosts. It allows the terminal to recognize a GPRS network and use IPv6.

In the operations and management area the collaboration between 3GPP and the IETF was in the development and modification of the COPS and DIA-METER protocols. A modified version of COPS was chosen to be used in the IMS called COPS-PR. The DIAMETER protocol is used as the base protocol and applications are developed on it. These applications and command codes are provided by the IETF in RFC 3589.

In the transport area most of the work done is for the development of the *session initiation protocol* (SIP). RFC 3261 was an outcome of this effort. However, this was not sufficient as the requirements of 3GPP were not being met. To accomplish this, IETF created a new working group called SIPPING, which collected and prioritized the SIP requirements and forwarded them to the SIP working group, where the actual work was done. This process of collaborative effort to develop SIP is documented in RFC 3427.

4.9 Indian Telecommunication Industry

In this section, a brief description is given of the current Indian telecommunication industry. The emerging trends of the industry are analyzed and the future trend is also presented. Some of the potential challenges that may act as obstacles to growth of this industry and the next generation network deployment in India are also discussed.

A brief presentation is also made on the current and the emerging trends in the Indian telecommunication industry. Particular focus is paid on the challenges that will be faced by the industry to sustain its growth rate and deploy next generation networks and services. Finally, various challenges and open issues in convergence and network interoperability are presented.

India is one of the fastest growing wireless markets in the world. It is now the second-largest telecom market, just after China. Despite the global economic slowdown, the Indian telecom industry continues to grow substantially, delivering strong returns on investment, fuelled by the growth in the wireless industry. The wireless subscriber base grew at a CAGR of 61% over FY04-FY09, while fixed-line subscribers dropped to 38 million in FY09 from 40.9 million in FY04. The subscriber base reached 392 million as of March 2009 with more than 10 million subscribers being added every month. However,

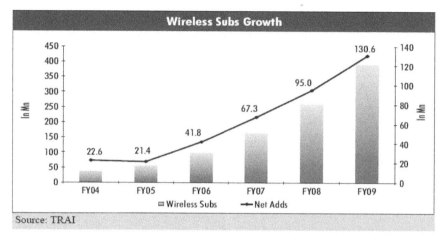

Figure 4.10 Wireless subscriber growth and net additions

India accounts for 7% of the total subscribers in the Asia Pacific. Thus, low mobile penetration provides huge growth potential [114].

Moreover, the teledensity levels between the urban and rural areas vary widely, which suggests untapped potential in the rural segment. It appears quite clearly that the rural areas will be the next growth driver for the Indian telecom industry. However, operators will face certain challenges such as high operating and maintenance costs and low ARPUs and MOU.

A continued decline in tariffs and disruptive entry pricing by new players pose risks to the margins of most of the companies. The wireless Average Revenue per User (ARPU) is expected to fall further due to increased competition. Also, incremental subscribers are now mainly from the rural areas with lower usage, which should reduce the elasticity of MOU with the decline in tariffs. Increased revenue is expected from non-voice services to support revenue growth.

4.9.1 Current Trends

Subscriber base: The Indian telecom market is characterized by low penetration and huge growth potential. The wireless subscriber base grew at a 61% CAGR over FY04-FY09 while fixed-line subscribers dropped to 38 million in FY09 from 40.9 million in FY04. Since wireless penetration is ~30% at present, there is a huge growth potential.

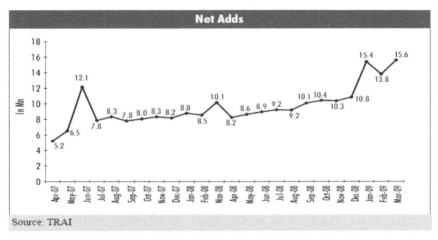

Figure 4.11 Monthly net additions of wireless subscribers

MOU growth not in line with fall in tariff: The Average Revenue per User (ARPU) for GSM operators dropped to INR220/month in December 2008 from INR316/month in December 2006, an average drop of 4.4%. Despite this, GSM operators' revenue grew an average 6.2% during the same period. This indicates that subscriber growth and higher Minutes of Use (MOU) per subscriber have more than offset the decline in ARPU. ARPU has been declining sharply as a result of intense competition among operators, fall in tariffs, regulatory policies such as phasing out of Access Deficit Charge (ADC), and reduction in termination charges. On the other hand, MOU has not increased in line with the fall in tariffs, mainly due to low usage of incremental subscribers.

Rapid growth in investments: The boom in the domestic telecom market has been attracting huge investments, which will likely accelerate with the launch of new services such as 3G and the entry of new players. The telecom industry saw 145% growth in FDI to INR108 billion over April-January 2009, compared with INR43.96 billion in the year-ago period (Source: DIPP). The pace of investments is expected to accelerate further with the entry of new license holders as well as the auction of 3G spectrum. Furthermore, buoyed by the rapid surge in the subscriber base, the domestic players too are planning huge investments for FY10.

Industry moving towards consolidation: The Indian telecom industry saw numerous mergers and acquisitions in the past. Many foreign players have already entered the industry by acquiring stakes in Indian players, who have acquired licenses recently. The 3G auctions would also lead to many foreign players entering the sector. However, as DOT has stipulated that for bidding for 3G spectrum an entity must hold a Universal Access Service (UAS) license, many foreign players might tie up with existing UAS licensees, which in turn could result in mergers.

4.9.2 Future Trends

Subscriber growth: The rural subscriber base has been growing substantially due to greater focus of the telecom service providers on rural operations. The rural subscribers accounted for almost 30% of the total subscriber base as of December 2008. However, there is an untapped potential in rural areas (where almost 70% of the population resides), as rural teledensity is quite low at 12.6% vs. urban teledensity of 81.3%. Quarterly rural subscriber additions have begun to exceed urban subscriber additions. Also, growth in rural subscribers has exceeded that in urban subscribers. Going forward, with a shift in focus towards rural areas, it is logical to believe that rural subscriber growth will continue to outpace urban subscriber growth.

Although rural areas would be the next growth driver, operators are expected to face certain challenges:

- *High network OPEX*: Operation and maintenance costs of cell sites in rural areas are high due to lower availability of electricity and thus operators are forced to depend on diesel for power supply. According to the FICCI-BDA wireless broadband report, cell site operating costs in rural areas are estimated to be 25% higher at $1,410/month vs. $1,050/month in urban areas.
- *Low ARPUs and MOU*: Due to lower per capita income and low usage, ARPUs and MOU are lower in rural areas than in urban areas.

Apart from huge Capital Expenses (CAPEX) and Operational Expense (OPEX) requirements, other constraints in increasing penetration in rural areas include acquisition of land, unavailability of cheap and fast backhaul connectivity, lack of continuous power supply, and low literacy levels. To increase rural penetration, TRAI has taken numerous initiatives such as recommending bringing mobile services under the ambit of USOF and sharing of infrastructure to receive support from USOF, and supporting backbone infrastructure through USOF. Therefore, low penetration, coupled with factors such as increasing affordability, lower handset prices and TRAI initiatives, suggests significant potential in rural areas. It is possible that the operators will expand coverage in rural areas to gain the first-mover advantage. This in turn should help improve rural teledensity.

Launch of new services: Services like IPTV and DTH services are being provided by some operators. For example, to leverage its existing fixed line infrastructure, Bharti and Airtel has started to offer IPTV and DTH services. While IPTV is limited initially to few cities, the company believes that DTH would become a mass product. Bharti will provide bundled services, commonly known as *triple play* – voice, broadband and TV services on one platform.

Declining ARPUs to put pressure on margin: The telecom industry is seeing adds of over 10 million subscribers every month, but the increasing base is not resulting in an increase in usage, as reflected by MOU. On the revenue front, growth is slowing down with a higher base and due to the consistent fall in ARPU. Going forward, a shift in the operators' focus towards low-usage rural subscribers would further contribute to a fall in ARPU. Furthermore, competitive pressure will likely heighten with new operators rolling out services, which in turn would lead to downward pressure on tariffs. TRAI recently reduced the termination charges for all domestic calls to INR 0.2/min from INR 0.3/min, which should further aid the tariff decline, pressurizing ARPUs. This, along with slow subscriber base growth, might result in slowing top-line growth, which, together with increasing competition and higher CAPEX to expand business, should put pressure on margins. However, to scale down the effect, the wireless operators are looking to increase value added service revenue similar to the developed markets, where operators have shifted focus towards data revenue, which has led to an improvement in margins.

Higher spectrum charges: The GSM operators were given an initial allotment of 4.4 MHz for network roll-outs, which was bundled along with the telecom license. However, the regulator has proposed a 1% increase in usage charges for spectrum up to 8 MHz and a 2% increase for spectrum up to and above 8 MHz. While a final decision is pending, this will likely impact majority of the players, as most of them have spectrum over 6.2 MHz in most of the circles.

Future growth expected in non-voice (data) revenue in highly penetrated areas: Circle-wise, penetration levels are high in metros and A Circles in terms of voice telephony. These circles are associated with high ARPUs. Therefore, with the voice telephony market nearing maturity in these areas, ARPUs are declining at a much faster pace. As ARPUs decline and voice becomes commoditized, the challenge for the telecom industry would be to retain customers, develop alternative revenue streams, and create a basis for differentiation in high-churn markets. This would in turn result in mobile operators shifting towards non-voice revenue, similar to the developed markets, where operators have shifted towards data revenue. It seems that data revenue will continue to offset the loss from voice ARPU. Moreover, it is rising as a percentage of wireless revenue. Globally, mobile data revenue growth is increasing and the voice market has been seeing a decline. In India as well, a similar trend is being observed, especially in the metros and urban areas, where mobile operators have shifted focus from customer acquisition to promoting value-added services. They are also looking at data as the next growth driver and a significant source of revenue. However, at present, data revenue contributes just 8–10% to the operators' overall revenue. Data revenue growth is led by SMS, with ring tones and caller ring-back tones being the next largest driver. Furthermore, 3G, which is expected to be the driving factor for high-end data services, should also help boost data revenue in India.

Infrastructure sharing: Recognizing the critical importance of infrastructure to offer wireless telecom services, TRAI has been taking various initiatives to encourage infrastructure sharing. TRAI considered the passive infrastructure sharing issue and recommended active infrastructure sharing and backhaul in April 2007. Passive infrastructure constitutes about 60% of the total cell site cost. Therefore, sharing would enable operators to reduce CAPEX to a large extent. Passive infrastructure sharing has provided a new business stream for telecom players and they appear to be realizing the benefits. Large players with a pan-Indian footprint have hived off their tower infrastructure units into companies to unlock value. For new players, who are expanding network, passive infrastructure sharing entails lower CAPEX and faster roll-out of services.

According to TRAI's estimates, India would need about 350,000 towers by 2010 to cater to a target subscriber base of 500 million. Due to strong demand for towers, given a rising subscriber base, tower companies have set aggressive target roll-outs. Many independent tower companies have also announced aggressive tower roll-outs to capture the growth opportunity. By 2010, Bharti Infratel plans to have 115,000 towers and Reliance plans to follow suit with 63,000 towers, GTL Infrastructure is also planning 25,000 towers. This is about 50% of the expected 350,000 towers.

Such aggressive tower roll-outs will likely result in an oversupply over a short period of time. However, the existing operators' expansion plans and the entry of new players should create demand for towers. It is likely that most of the new players will lease towers to ensure faster roll-out of networks rather than incur CAPEX. Further, with large number of independent tower companies and integrated telecom players in the industry, one can only expect consolidation in the tower industry, signs of which are already visible.

4.9.3 Challenges

Some of the challenges for telecommunication industry in India are: regulatory uncertainties, severe competition, lack of affordability for the rural consumers, infrastructure and social issues. These challenges are briefly discussed below. The two biggest external challenges for telecommunication industry in India are regulatory uncertainty and competition.

Regulatory Uncertainty

Lack of a clear regulatory roadmap has been one of the major challenges for the Indian telecom industry. Uncertainty about the policy and regulatory framework revolve around the following:

- *Final decision on spectrum not yet taken*: With TRAI and TEC increasing the subscriber base criteria, it has now become difficult for the existing operators to get additional spectrum. For efficient and quality services, spectrum availability is a must. As these norms for spectrum allocation are implemented, the existing operators will have to increase their subscriber base to get additional spectrum. However, a final decision has not been taken yet. The scarcity of spectrum is an industry-wide issue, which needs to be addressed. It may affect smaller players more because huge CAPEX is required to upgrade the network due to the spectrum crunch. The extent to which an operator would be affected would depend on its capability to maintain the quality of its services, which becomes even more important in the wake of the introduction of mobile number portability (MNP). From a subscriber perspective, it seems that net additions will not be affected much with the spectrum crunch. The criteria for subscribing a mobile connection do not include determining the amount of spectrum with a service provider. However, the existing subscribers may switch over to other providers due to deterioration in QoS. With the launch of MNP, switching over to another service provider would be easier. Another impact would be increased competition between service providers. This may result in aggressive pricing strategies to retain subscribers. Spectrum scarcity is a fundamental problem. Not much of spectrum is available for commercial cellular services despite most service operators having already become eligible for additional spectrum. Moreover, the Defense Ministry has delayed vacating of spectrum for a long time now.

- *Delay in 3G auctions*: India has been waiting for the launch of 3G services for a long time now. Although the government has announced a comprehensive policy for 3G, it has been surrounded by controversies, leading to a series of amendments. Procedural delays and disagreements between the Finance Ministry, DoT, and TRAI on the reserve price for spectrum auctions have been delaying the schedule. The government has not been able to lay down a clear roadmap for 3G auctions. Furthermore, the delay in auctions is benefiting state-owned companies such as BSNL and MTNL, as spectrum has been allocated to them, whereas private players will have to wait for the auctions to begin.

- *Uncertainty about defense vacating the spectrum*: The signing of an MOU between the DoT and the Defense Ministry for vacating spectrum has been delayed. The Defense Ministry earlier agreed to release 10 MHz of spectrum for third generation mobile services and another 5 MHz for existing GSM operators immediately after signing the agreement. The DoT, on its part, agreed to lay an optical fiber cable for the armed forces, connecting over 270 locations over three years. However, with the Defense ministry now proposing new conditions in the MoU, the vacation of spectrum has been delayed. These conditions include powers to take back the vacated spectrum if DoT fails to roll out an alternative optical fiber cable network for the armed forces within the stipulated period. It has also proposed that spectrum should be reverted to the armed forces in case operators do not start using it within a specified time. The DoT has so far not agreed to these conditions, which creates uncertainty about spectrum availability for mobile operators. Thus, an increasing delay in signing of the MoU could put off the 3G auctions further and also increase congestion in existing cellular networks.

- *Uncertainty about introduction of MNP*: In August 2008, DoT announced the guidelines for implementation of intra-circle MNP with a timeline of mid-09 for metros and end-09 for the whole country. The DoT divided the country into two geographic zones, each of which would be handled by a different MNP provider. Each zone was further broken down into 11 service areas that represent cities within the zone. Zone 1 will cover the northern and western regions, while Zone 2 will include the south and east. The centralized operators for implementing MNP were finalized in March. Syniverse will cover northern and western states, while MNP Interconnection Telecom Solutions India, a joint venture of Telcordia, will cover the southern and eastern states. However, these centralized operators are required to roll out MNP in metros and A-category Circles within six months of receiving the license, and extend it to the other circles in the next six months. Therefore, there is uncertainty about the implementation of MNP across the country by December 2009.

Competition

To take advantage of the boom in the telecom sector, many new aspirants applied for licenses in the past. The DOT received as many as 575 applications. However, letters of intent were given to only those who applied before the due date. Apart from leading global telecom giants, those with presence in real estate and infrastructure sectors applied for licenses to be a part of the growth story. The DOT issued letters of intent to nine different companies. Of them, only two-Unitech and STel-were completely new. Others were Tata Teleservices, Loop Telecom (owned by BPL), Spice Communication, Shyam Telelink, Swan Telecom, Datacom and Idea Cellular. These letters of intent have now been converted into licenses. With the issue of start-up spectrum allocated to the new licensees, the number of operators in a particular Circle would be more than four. However, the entry of new players would not pose any threat to existing operators, as the chances of them affecting the market in the near term are low due to significant market and execution challenges. Apart from spectrum constraints, these players would face financing risks and severe competition, as they are late entrants and are likely to enter simultaneously in an already crowded market.

Lack of affordability: When considering any technology for rural India, the issue of affordability must be addressed first. Given the income level, one must determine the cost of any sustainable solution. It is reasonable to expect an expenditure on telecommunication services of only around US$1.5 per month on the average (2% of household income). In this scenario, mobile handsets with advanced features will be too expensive and beyond afford for the majority of the rural population.

Lack of infrastructure: In many villages in India uninterrupted supply of electricity is not available. This causes a hindrance to the spread of Internet in rural India. Moreover, rural India suffers from poor availability of backhaul connection. Finally, the last mile in rural India does not exist. The cost of laying down last mile is prohibitively expensive.

Social issues: For many rural families in India, a phone is not a personal object. Mobile phones, like a land line, are used by the entire family. Unless this attitude changes the rural teledensity will remain low.

4.10 Research Challenges

The design of next generation converged networks will pose a number of significant research challenges. There are several crucial issues that need to be investigated before a true convergence can happen in the network and service perspective. Some of the important issues are enumerated below:

- *Architectural Framework*: The migration towards NGN changes the architecture and topology of networks which potentially involves several structural changes, such as a reorganization of core network nodes and changes in the number of network hierarchy levels. The shift to IP networks also raises questions whether interconnection frameworks need to be revised.
- *QoS issues*: NGN will be all-IP based and will have to provide guaranteed QoS to mobile terminals. QoS provisioning in a heterogeneous wireless and mobile networks will bring in new problems to mobility management, such as location management for efficient access and timely service delivery, QoS negotiation during inter-system handoff, etc.
- *Design of user terminals*: The design of single user terminal that is able to autonomously operate in different heterogeneous access networks will be another important research challenge. This terminal will have to exploit various surrounding information (e.g., communication with localization systems, cross-layer with network entities etc.) in order to provide richer user services (e.g., location/situation/context-aware multimedia services). This will also put strong emphasis on the concept of cognitive radio and cognitive algorithms for terminal re-configurability.
- *Location and handoff management in wireless overlay networks*: Future wireless networks will be inherently hierarchical where access networks have different coverage areas. Mobility management in wireless overlay networks will pose a difficult challenge to solve.
- *Cross-layer optimization*: Design of efficient cross-layer-based approaches will be instrumental in developing new mobility management schemes. It has already been observed through research that cooperation between the network and link layers is able to improve the performance

of mobility management in IP-based heterogeneous communication environment. Information from the link layer such as signal strength and velocity of mobile terminals may help the decision making of mobility management techniques at the network layer. In cross-layer optimization, how to cooperate, how tight the cooperation is, and how much information is to be exchanged between the two layers are possible research issues.

- *Other issues*: Efficient use of spectrum, fault-tolerance, availability of network services, enhanced security, intelligent packet and call routing, intelligent gateway discovery and selection protocol design and development of a unified protocol stack and vertical protocol integration mechanisms are some of the other important research issues in next-generation heterogeneous networks.

4.11 Conclusions

This chapter has provided a clear picture of a converged, all IP communications environment, which fulfils almost all the expectations and requirements of a NGN system. The aim was to help the reader comprehend the concept of convergence, its drivers and enablers in NGN. Various issues of NGN are discussed and the current and future trends of standardization activities for NGN are presented in detail. IMS is depicted as a major enabler for achieving convergence. It presented a case study to illustrate how IMS can be used to achieve convergence in a residential networking environment. Some of the applications mentioned may be a bit advanced and not feasible from the perspective of India, but they are certainly going to be deployed in the near future. The emerging trends of the Indian telecommunication industry are also discussed and some of the challenges that the industry is facing today for its growth and evolution are presented. Finally some open problems in the domain of NGN and convergence are discussed for the potential researchers to explore.

The chapter also focuses on issues of convergence of services like voice, video and data in the next generation networks and how it can lead to more competition in individual markets for each of these services. Convergence of services puts increasing competitive pressure on mobile carriers from the IP world. However, the trend towards horizontal integration of infrastructures, market and services could lead to strengthening of market power as there may be relatively few companies that can package voice, video and data services in a single bundled offer to end users.

References

[1] http://www.openhandsetalliance.com/index.html.
[2] OECD, Participative Web: User Created Content, http://www.oecd.org/sti/interneteconomy.
[3] http://www.oecd.org/sti/interneteconomy.
[4] http://www.slingbox.com.
[5] ITU-T Recommendation Y.2001, approved in December 2004, http://www.itu.int/rec/TRECY2001-200412-I/en.
[6] V. Reding, European Commissioner for Information Society and Media, The access revolution: An evolution of regulation for competition, Address to the Annual KPN Event, Brussels, 14 January 2008.
[7] DOCSIS: Data over cable service interface specifications, http://docsis.org.
[8] OECD Communications Outlook 2007; OECD, The implications of WiMAX for competition and regulation. OECD Universal Access and NGN, 2006.
[9] OECD, The implications of WiMAX for competition and regulation, http://www.oecd.org/LongAbstract/0.3425.en_2649_34223_36218740_1_1_1_1.00.html.
[10] OFCOM, Regulatory challenges posed by next generation access networks, http://www.ofcom.org.uk/research/telecoms/reports/nga/nga.pdf, 23 November 2006.
[11] Telecom Italia Project NGN2, Presentation at the meeting with financial community, http://www.telecomitalia.it/TIPortale/docs/investor/ID_Pileri_NT_OK_noNote.pdf, March 2007.
[12] The IEEE P1901 Work Group for Broadband over Powerline, http://grouper.iee.org/groups/1901.
[13] OECD Paper on fixed-mobile convergence: Market developments and policy issues, http://www.oecd.org/sti/ict.
[14] FCC Annual Report and Analysis of Competitive Market Conditions with Respect to Domestic and International Satellite Communications Services, March 2007.
[15] E. Langmantel, NGN as multimedia implementation of legacy Telco model. Presentation at the ITU NGN Workshop, Geneva, May 2006 and ERG Report on IP Interconnection, http://erg.eu.int/documents/cons/index_en.htm, February 2007.
[16] OECD Internet Traffic Prioritization, DSTI/ICCP/TISP(2006)4/FINAL, Paris, April 2007.
[17] ITU Rec. Y.2001, 2004.
[18] ERG Interconnection Report, 2007.
[19] S.J. Marcus, Interconnection on an IP-based NGN environment. Discussion Paper for the ITU Global Symposium for Regulators, Dubai, http://www.itu.int/ITUD/treg/Events/Seminars/GSR/GSR07/agenda-documents.html, February 2007.
[20] F. Fuentes, Presentation, European Telecommunication Platform, http://www.etp-online.org.
[21] OECD Fixed-Mobile Convergence: Market Developments and Policy Issues, page 8, Figure 1, 2007 http://www.oecd.org/dataoecd/20/26/38309911.pdf (last accessed April 2007).

[22] R. Ruggiero, TI CEO. Telecom Italia, presentation, `http://www.telecomitalia.com/analystmeeting2007/eng/index.html`.

[23] OECD Study on Participative Web and User Created Content. ERG Workshop on Next Generation Networks: "NGN Regulation and Investment", Turin, 17 April 2007.

[24] IMS Release 7, `http://www.etsi.org/tispan/`.

[25] A. Rossi, Italtel, Technical progress, market evolution and the regulation of the electronic communications sector in the EU, Paul Richards, BT, March 2007.

[26] `http://www.techabulary.com/ims/ims_ip_multimedia\_subsytem.pdf`.

[27] R. Cuel, D. Ponte, and A. Rossi, Towards an Open/Web2.0 scientific publishing industry? Preliminary findings and open issues. `http://wiki.liquidpub.org/mediawiki/upload/b/b3/CuelPonteRossi09.pdf`.

[28] `http://www.ericsson.com/products`.

[29] `http://www.3gpp.org`.

[30] 3GPP Active Work Program, Voice call continuity (VCC) between CS and IMS (incl. I-WLAN), `http://www.3gpp.org/ftp/Specs/html-info/FeatureOrStudyItemFile-32091.htm`.

[31] `http://www.umatoday.com/umaOverview.php`.

[32] Summary Report of the OECD Foresight Forum, Next generation networks: Evolution and policy considerations, Budapest, `http://www.oecd.org/sti/ict`, 3 October.

[33] ECC Report 27, Next generation network developments and their implications for the new regulatory regime, October 2003.

[34] OECD Internet Traffic Exchange, Market developments and measurement of growth, OECD, Paris. `http://www.oecd.org/document/27/0.3343.en_2649_34225_25496027_1_1_1_1.00.html`, April 2006.

[35] `http://www.arcep.fr`. A study conducted by ARCEP.

[36] Competition in the Internet Access Market: Current Regulatory Issues, `http://www.regtp.de/en/aktuelles/reden/02713/03/index.html`.

[37] ERG Final Report on IP Interconnection. `http://erg.eu.int/doc/publications/erg_07_09_rept_on_ip_interconn.pdf`.

[38] `http://www.supra.net.nz`.

[39] ITU-T Recommendation E.164.

[40] OECD Foresight Forum, Next generation networks: Evolution and policy considerations, held in Budapest, October 2006, presentation, `http://www.oecd.org/dataoecd/24/5/38079155.pdf`.

[41] IPv6 Transition and Co-existence with Ipv4, `http://www.6diss.org/workshops/ca/integration.pdf`.

[42] The TCP/IP Guide-TCP/IP Protocol Suite and Architecture, `http://www.tcpipguide.com`.

[43] Session Initiation Protocol (RFC 3261), `http://www.ietf.org/rfc/rfc3261.txt`.

[44] Securing VoIP Networks. `http://www.voipplanet.com/backgrounders/article.php/11924_3707506_2`.

[45] `http://www.itu.int/osg/spu/infocom/enum`.

[46] `http://www.ietf.org`.

[47] OECD Rethinking Universal Service for a Next generation Network Environment, DSTI/ICCP/TISP(2005)5/FINAL.

[48] H. Williams, Universal service obligation in a competitive environment.

[49] Directive 2002/22/EC of the European Parliament and of the Council on Universal Service and Users' Rights Relating to Electronic Communications Networks and Services, Official Journal L 108, April 2002.

[50] OFCOM Universal Service Review, `http://www.ofcom.org.uk/consult/condocs/uso/main`.

[51] 47 US Code 254(c) (1).

[52] The Implications of WiMAX for Competition and Regulation, DSTI/IIP/TISP (2005)4/Final Report of the Working Party on Telecommunication and Information Services Policies.

[53] EU Council Resolution on Law Enforcement Operational Needs with Respect to Public Telecommunication Networks and Services, `http//cryptome.org/eu-intercept.htm`.

[54] `http://www.parlamento.it/leggi/051551.htm`.

[55] R. Foster, Future broadcasting regulation. Report commissioned by the UK Department for Culture, Media and Sport.

[56] OECD, The Spectrum Dividend: Spectrum Management Issues, DSTI/ICCP/TISCP (2006) 2/FINAL.

[57] `http://www.joost.com`.

[58] T. Gardam and D.A.L. Levy, The price of plurality – Choice, diversity and broadcasting institutions in the digital age, 2008.

[59] EU Directive on Television without Frontiers (TVWF).

[60] R. Foster, Public interest content and the role of the PSP, `http://www.ofcom.org.uk/consult/condocs/c4publicinterest.pdf`.

[61] R. Barvainis, Spectrum management issues, `http://www.hiaper.ucar.edu/archive/02workshop/barvainis.pdf`.

[62] Industry Canada's Report Study of Market-based Exclusive Spectrum Rights, `http://www.ic.gc.ca/eic/site/smt-gst.nsf/eng/sf09401.html`.

[63] G. Ingram, OECD NGN Technical Foresight Forum, NGN Security, `http://www.oecd.org/document/12/0.2340.en_2649_33703_33703_37392780_1_1_1_1_1.00.html`.

[64] Wikipedia, Security through obscurity, `http://en.wikipedia.org/wiki/Security_through_obscurity`.

[65] OECD Workshop Social and Economic Factors Shaping the Future of the Internet, `http://www.oecd.org/sti/ict`.

[66] OECD The development of policies for the protection of critical information infrastructures (CII).

[67] G. Galler, European Commission, DG Information Society and Media, Presentation at the ETSI Meeting: A Strategy for a Secure Information Society: Dialogue, Partnership and Empowerment, 2007.

[68] ITU-T Recommendation Y.2201 NGN Release 1 requirements, and ITU-T Recommendation Y.2012 Functional requirement and architecture and architecture of the NGN.

[69] ITU-T Recommendation Y.2701, Security requirements for NGN Release 1.

[70] B. Gamm, B. Howard, and O. Paridaens, Security features required in an NGN, `http://sysdoc.doors.ch/ALCATEL/10GammaGB.pdf`.

[71] Alcatel-Lucent paper on Security 3GPPIMS to TISPAN NGN, `http://www.alcatellucent.com/com/en/appcontent/apl/S0512-TISPAN_NGN-EN-tcm172-521381635.pdf`.

[72] IDA Singapore Report, Reference specification for next generation networks: Technical framework, 2007.

[73] OECD Working Party on Security and Privacy, IDM resources. `www.oecd.org/sti/securityprivacy/idm`.

[74] M. Euchner, Filling the gaps of IDM in third and in next generation networks: Standardized network-centric IDM as an enabler for secure applications, Presentation in OASIS Identity and Trusted Infrastructure Workshop: Evolutionary Milestones, Barcelona, October 2007.

[75] 3GPP, Service requirements for the Internet Protocol (IP) multimedia core network subsystem; Stage 1, Technical Specification Group Services and System Aspects, 3rd Generation Partnership Project (3GPP), TS 22.228 V7.3.0, `http://www.3gpp.org/ftp/specs/latest/Rel-7/22_series/`, December 2005.

[76] G. Camarillo and M.-A. Garcia-Martin, *The 3G IP Multimedia Subsystem (IMS): Merging the Internet and the Cellular Worlds*. John Wiley and Sons, 2004.

[77] 3GPP, End-to-end Quality of Service (QoS) concept and architecture, Technical Specification Group Core Network, 3rd Generation Partnership Project (3GPP), TS 23.207 V6.6.0, `http://www.3gpp.org/ftp/specs/latest/Rel-6/23_series/`, September 2005.

[78] P. Calhoun, J. Loughney, E. Guttman, G. Zorn, and J. Arkko, Diameter base protocol, RFC 3588, Internet Engineering Task Force (IETF), `ftp://ftp.rfc-editor.org/in-notes/rfc3588.txt`, September 2003.

[79] J. Rosenberg, H. Schulzrinne, G. Camarillo, A. Johnston, J. Peterson, R. Sparks, M. Handley, and E. Schooler, SIP: Session Initiation Protocol, RFC 3261, Internet Engineering Task Force (IETF), `ftp://ftp.rfc-editor.org/in-notes/rfc3261.txt`, June 2002.

[80] 3GPP, IP Multimedia Subsystem (IMS); Stage 2, Technical Specification Group Services and System Aspects, 3rd Generation Partnership Project (3GPP), TS 23.228 V7.2.0, `http://www.3gpp.org/ftp/specs/latest/Rel-7/23_series/`, September 2005.

[81] G. Camarillo, W. Marshall, and J. Rosenberg (Eds.), Integration of Resource Management and Session Initiation Protocol (SIP), RFC 3312, Internet Engineering Task Force (IETF), `ftp://ftp.rfc-editor.org/in-notes/rfc3312.txt`, October 2002.

[82] 3GPP, 3G security; Access security for IP-based services, Technical Specification Group Services and System Aspects, 3rd Generation Partnership Project (3GPP), TS 33.203 V7.0.0, `http://www.3gpp.org/ftp/specs/latest/Rel-7/33_series/`, December 2005.

[83] 3GPP, *End-to-end Quality of Service (QoS) concept and architecture*, Technical Specification Group Core Network, 3rd Generation Partnership Project (3GPP), TS 23.207 V6.6.0, `http://www.3gpp.org/ftp/specs/latest/Rel-6/23_series/`, September 2005.

[84] J. Rosenberg and H. Schulzrinne, Session Initiation Protocol (SIP): Locating SIP Servers, RFC 3263, Internet Engineering Task Force (IETF), `ftp://ftp.rfc-editor.org/in-notes/rfc3263.txt`, June 2002.

[85] 3GPP, Open Service Access (OSA); Application Programming Interface (API); Part 1: Overview, Technical Specification Group Core Network, 3rd Generation Partnership Project (3GPP), TS 29.198-1 V5.8.2, http://www.3gpp.org/ftp/specs/latest/Rel-5/29_series/, December 2005.

[86] 3GPP, Customized Applications for Mobile network Enhanced Logic (CAMEL) Phase 4; CAMEL Application Part (CAP) specification for IP Multimedia Subsystems (IMS), Technical Specification Group Core Network, 3rd Generation Partnership Project (3GPP), TS 29.278 V7.0.0, http://www.3gpp.org/ftp/specs/latest/Rel-7/29_series/, December 2005.

[87] 3GPP, Mobile Application Part (MAP) specification, Technical Specification Group Core Network and Terminals, 3rd Generation Partnership Project (3GPP), TS 29.002 V7.2.0, http://www.3gpp.org/ftp/specs/latest/Rel-7/29_series/, December 2005.

[88] ITU-T, *Signalling System No. 7 – ISDN User Part functional description*, Recommendation Q.761, International Telecommunication Union (ITU), http://www.itu.int/rec/recommendation.asp?type=folders\lang=e\parent=T-REC-Q.761, December 1999.

[89] ITU-T, Bearer Independent Call Control Protocol, Recommendation Q.1901, International Telecommunication Union (ITU), http://www.itu.int/rec/recommendation.asp?type=folders\lang=e\parent=T-REC-Q.1901, June 2006.

[90] ITU-T, Gateway Control Protocol, Recommendation H.248, International Telecommunication Union (ITU), http://www.itu.int/rec/recommendation.asp?type=folders\lang=e\parent=TREC-H.248, June 2000.

[91] ITU-T, Functional description of the message transfer part (MTP) of Signalling System No. 7, Recommendation Q.701, International Telecommunication Union (ITU), http://www.itu.int/rec/recommendation.asp?type=folders\lang=e\parent=T-REC-Q.701, April 1993.

[92] R. Stewart, Q. Xie, K. Morneault, C. Sharp, H. Schwarzbauer, T. Taylor, I. Rytina, M. Kalla, L. Zhang, and V. Paxson, Stream Control Transmission Protocol, RFC 2960, Internet Engineering Task Force (IETF), ftp://ftp.rfc-editor.org/in-notes/rfc2960.txt, October 2000.

[93] H. Schulzrinne, S. Casner, R. Frederick, and V. Jacobson, RTP: A Transport Protocol for Real-Time Applications, RFC 3550/STD 0064, Internet Engineering Task Force (IETF), ftp://ftp.rfc-editor.org/in-notes/rfc3550.txt, July 2003.

[94] C. Rigney, S. Willens, A. Rubens, and W. Simpson, Remote Authentication Dial in User Service (RADIUS), RFC 2865, Internet Engineering Task Force (IETF), ftp://ftp.rfceditor.org/in-notes/rfc2865.txt, June 2000.

[95] R. Braden, L. Zhang, S. Berson, S. Herzog, and S. Jamin (Eds.), Resource ReSerVation Protocol (RSVP) – Version 1 Functional Specification, RFC 2205, Internet Engineering Task Force (IETF), ftp://ftp.rfc-editor.org/in-notes/rfc2205.txt, September 1997.

[96] S. Herzog, RSVP Extensions for Policy Control, RFC 2750, Internet Engineering Task Force (IETF), ftp://ftp.rfc-editor.org/in-notes/rfc2750.txt, January 2000.

[97] K. Kompella and J. Lang, Procedures for Modifying the Resource reSerVation Protocol (RSVP), RFC 3936/BCP0096, Internet Engineering Task Force (IETF), `ftp://ftp.rfc-editor.org/innotes/rfc3936.txt`, October 2004.

[98] S. Blake, D. Black, M. Carlson, E. Davies, Z. Wang, and W. Weiss, An Architecture for Differentiated Service, RFC 2475, Internet Engineering Task Force (IETF), `ftp://ftp.rfceditor.org/in-notes/rfc2475.txt`, December 1998.

[99] D. Grossman, New terminology and clarifications for Diffserv, RFC 3260, Internet Engineering Task Force (IETF), `ftp://ftp.rfc-editor.org/in-notes/rfc3260.txt`, April 2002.

[100] S. Kent and R. Atkinson, Security architecture for the Internet protocol, RFC 2401, Internet Engineering Task Force (IETF), `ftp://ftp.rfc-editor.org/in-notes/rfc2401.txt`, November 1998.

[101] 3GPP, 3G security; Access security for IP-based services, Technical Specification Group Services and System Aspects, 3rd Generation Partnership Project (3GPP), TS 33.203 V7.0.0, `http://www.3gpp.org/ftp/specs/latest/Rel-7/33_series/`, December 2005.

[102] S. Kent and R. Atkinson, IP Encapsulating Security Payload (ESP), RFC 2406, Internet Engineering Task Force (IETF), `ftp://ftp.rfc-editor.org/in-notes/rfc2406.txt`, November 1998.

[103] D. Harkins and D. Carrel, The Internet Key Exchange (IKE), RFC 2409, Internet Engineering Task Force (IETF), `ftp://ftp.rfc-editor.org/in-notes/rfc2409.txt`, November 1998.

[104] 3GPP, 3G Security; Network Domain Security; IP network layer security, Technical Specification Group Services and System Aspects, 3rd Generation Partnership Project (3GPP), TS 33.210 V7.0.0, `http://www.3gpp.org/ftp/specs/latest/Rel-7/33_series/`, December 2005.

[105] D. Durham, J. Boyle, R. Cohen, S. Herzog, R. Rajan, A. Sastry (Eds.), The COPS (Common Open Policy Service) Protocol, RFC 2748, Internet Engineering Task Force (IETF), `ftp://ftp.rfceditor.org/in-notes/rfc2748.txt`, January 2000.

[106] J. Walker and A. Kulkarni (Eds.), Common Open Policy Service (COPS) Over Transport Layer Security (TLS), RFC 4261, Internet Engineering Task Force (IETF), `ftp://ftp.rfceditor.org/in-notes/rfc4261.txt`, December 2005.

[107] 3GPP, Access to network application functions using Hypertext Transfer Protocol over Transport Layer Security (HTTPS), Technical Specification Group Services and System Aspects, Generic Authentication Architecture (GAA), 3rd Generation Partnership Project (3GPP), TS 33.222 V7.0.0, `http://www.3gpp.org/ftp/specs/latest/Rel-7/33_series/`, September 2005.

[108] ATIS NGNFG, ATIS Next Generation Network (NGN) Framework, Part I: NGN Definitions, Requirements, and Architecture, Next Generation Network Focus Group, Technology and Operations Council (TOPS), Alliance for Telecommunications Industry Solutions (ATIS), Issue 1.0, `http://www.atis.org/topsc/Docs/ATIS-NGN-Framework-Part1-Issue1.pdf`, November 2004.

[109] ATIS NGNFG, Next Generation Network (NGN) Framework, Part II: NGN Roadmap 2005, Next Generation Network Focus Group, Technology and Operations Council (TOPS), Alliance for Telecommunications Industry Solutions (ATIS),

Issue 1.0, http://www.atis.org/topsc/Docs/ATIS_Framework_Part_2_NGN_Roadmap_FINAL.pdf, August 2005.

[110] H. Alvestrand, A mission statement for the IETF, RFC 3935/BCP 0095, Internet Engineering Task Force (IETF), ftp://ftp.rfc-editor.org/in-notes/rfc3935.txt, October 2004.

[111] W. Marshall (Ed.), Private Session Initiation Protocol (SIP) extensions for media authorization, RFC 3313, Internet Engineering Task Force (IETF), ftp://ftp.rfc-editor.org/in-notes/rfc3313.txt, January 2003.

[112] S. Bradner, P. Calhoun, H. Cuschieri, S. Dennett, G. Flynn, M. Lipford, and M. McPheters, 3GPP2-IETF Standardization Collaboration, RFC 3131, Internet Engineering Task Force (IETF), ftp://ftp.rfc-editor.org/in-notes/rfc3131.txt, June 2001.

[113] J. Arkko, G. Kuijpers, H. Soliman, J. Loughney, and J. Wiljakka, Internet Protocol Version 6 (IPv6) for Some Second and Third Generation Cellular Hosts, RFC 3316, Internet Engineering Task Force (IETF), ftp://ftp.rfc-editor.org/in-notes/rfc3316.txt, April 2003.

[114] Research Report, Telecom Sector Rural Penetration drives Growth, http://www.moneycontrol.com/AistCMehta_Bharti_TataComm.pdf, May 2009.

[115] K. Knighston, Industry Canada, ITU NGN Architecture. Presentation at the ITU-IETF Workshop on NGN, May 2005.

[116] J. Horrocks, NGN and convergence models, myths, and muddle. OECD NGN Foresight Forum, 3 October 2006.

[117] ATIS Next Generation Network (NGN) Framework, Part I: Definitions, Requirements, and Architecture, Issue 1.0, p. 11, November 2004.

Biographies

Jaydip Sen obtained his Bachelor of Engineering (B.E) with honors in Electrical Engineering from Jadavpur University, Kolkata, India in 1993, his Master of Technology (M.Tech) with honors in Computer Science from Indian Statistical Institute, Kolkata, India in 2001, and his PhD from the Indian Institute of Technology, Kharagpur, India in 2007. Currently, he is associated with Tata Consultancy Services (TCS) Ltd., in the Wireless and Multimedia Innovation Lab, Kolkata, India. Prior to joining TCS, he had worked with Akamai Technologies Ltd., Bangalore, where he was involved in design and implementation of a number of communication protocols for wireless and wired networks. He has diverse experience in the field of communication and computing. His research areas include security in wired and wireless networks, intrusion detection systems, secure routing protocols in wireless ad hoc and sensor networks, secure multicast and broadcast communication in next generation broadband wireless networks, trust and reputation based systems, quality of service in multimedia communication in wireless networks, cross layer optimization based resource allocation algorithms in next generation wireless networks and privacy issues in ubiquitous computing. He has published more than 60 papers in reputed international journals and referred conference proceedings. He is a member of ACM and IEEE and also a working member of IEEE 802.16 group.

Munir Bashir Sayyad holds a degree in Electronics & Telecommunication Engineering from Swami Ramanand Tirth University Nanded (India) and followed post-graduate studies in Electronics & Telecommunication Engineering at Dr. Babasaheb Ambedkar Technological University Lonere (India). He has six years of experience as Faculty of Engineering & Technology. He has worked as Test & Verification Engineer in VLSI industry for 1.5 year. The last three years he has been working at Reliance Technology Innovation & Interoperability Test Labs, Reliance Communication, Navi Mumbai. Currently he is involved with the Project of IMS, WiMax and LTE Validation. His topics of interest are Signaling Protocol Analysis, IP Multimedia Subsystem, Web 2.0, WiMax, LTE, and Internet of Things.

Basavarj Hooli is M. Tech. (Computer Science and Engineering), IIT Mumbai, B.E. (Electronics and Communication Engineering) Karnatak University. He has over 23 years of experience in the IT industry with strong skills in execution, business development, strategies, innovation and thought leadership. He currently is with MphasiS Ltd, Pune where he is responsible for managing Telecom Clients from the delivery perspective for the past three and half years. He has proficiencies and expertise in Strong Telecom Domain with experience of working in all the TMN layers viz. Element Management, Network Management, Service Management and Business Management, development of telecom devices (Embedded Systems), experience in managing the legacy and midrange applications in Telecom space and managing their transformation projects, conceptualization and development of new products.

5

Internet of Things

P. Balamuralidhar[1], Santosh Bothe[2] and Prateep Misra[3]

[1]*TCS, Bangalore, India*
[2]*Sinhgad Institute of Technology, Lonavala*
[3]*Tata Consultancy Services Ltd.*

> The Internet of the future will be suffused with software,
> information, data archives, and populated with devices, appliances, and
> people who are interacting with and through this rich fabric.
>
> Vinton Cerf, one of the Founders of Internet

The Internet and World-Wide Web have been major drivers of globalization and have promoted the convergence of electronic communications and media services. Internet is continuing to become more pervasive, with the advent of low cost wireless broadband connectivity, by connecting to new embedded devices and handhelds. Further, this evolution will continue to emerge as an "Internet of Things (IoT)" where the web will provide a medium for physical world objects to take part in interaction. This way the digital information technology can integrate the physical world to the online world to provide a common interaction platform.

The number of world Internet users has grown 20-fold in the past decade to about 1.5 billion people in 2008, with the number of computer servers rising from 22.5 million to 489 million. Eventually, the entire world will have access through a variety of smart devices to the services that are available on the Internet. As we observe, the earth has 6.6 billion people, 50 billion machines and 50,000 billion things and the scope of internetworking them is too huge to say that the concept of the IoT is still a vision. However we see the use of current technologies to address a subset of high impact applications keeping a scalable architecture towards the giant vision. The movement from

R. Prasad (ed.), Future Trends and Challenges for ICT Standardization, 193–227.
© 2010 *River Publishers. All rights reserved.*

today's Internet of data to Internet of people and then Internet of things is evolutionary progressive and challenging.

This chapter is organized as follows. Section 5.1 gives an overview, vision and future applications of IoT. Section 5.2 outlines the architectural requirements of IoT with the focus on issues and challenges. This section also focuses on EPCGlobal network architecture in detail. Section 5.4 is about ubiquitous computing with the comparative study of wireless protocols and standards. In particular, this area has a main challenge in managing heterogeneity, QoS requirements and application specific Standardisation. This section also highlights radio spectrum for wireless connectivity Section 5.5 describes the mechanism for managing identity in a ubiquitous computing environment, use of IPV6 for identity management and host identity protocol for object identification. This section gives an account of the security and interoperability requirements for security. Section 5.6 describes the importance and challenges related to RFID, IoT and EPCGlobal networks with the challenges for standards developement. Section 5.7 briefly outlines the way ahead and Section 5.8 presents conclusions.

5.1 Overview

In a world that seems to move faster and faster developments are rapidly under way to take an important technological step into the wireless era, by embedding short-range mobile transceivers into a wide array of additional gadgets and everyday items, enabling new forms of communication between people and things, and between things themselves. A new dimension has been added to the world of information and communication technologies: from anytime, anyplace connectivity for anyone, we will now have connectivity for anything Connections will multiply and create an entirely new dynamic network of networks – an Internet of Things (IoT). The Internet of Things is neither science fiction nor industry hype, but is based on solid technological advances and visions of network ubiquity that are zealously being realized.

The IoT is a technological revolution that represents the future of computing and communications, and its development depends on dynamic technical innovation in a number of important fields, from wireless sensors to nanotechnology.

Considering the functionality and identity as central it is reasonable to define the IoT as "Things having identities and virtual personalities operating in smart spaces using intelligent interfaces to connect and communicate within social, environmental, and user contexts". A different definition, that puts the focus on the seamless integration, could be formulated as "Interconnected objects having an active role in what might be called the Future Internet" [1].

5.1.1 Internet of Things – Vision

The Internet of Things (IoT) describes a worldwide network of intercommunicating devices. It integrates ubiquitous communications, pervasive computing, and ambient intelligence. As pointed out earlier IoT must be seen as a vision where "things", especially everyday objects, such as home appliances, furniture, clothes, vehicles, roads and smart materials, and more, are readable, recognizable, locatable, addressable and/or controllable via the Internet. This will provide the basis for many new applications, such as energy monitoring, transport safety systems or building security. This vision may change with time, especially as synergies between Identification Technologies, Wireless Sensor Networks, Intelligent Devices and Nanotechnology will enable a number of advanced applications. Innovative use of technologies such as RFID, NFC, ZigBee and Bluetooth are contributing to create a value proposition for Internet of Things stakeholders. It also requires a smart integration of multiple enabling technologies as depicted in Figure 5.1.

A global communications platform needs to be developed to realize the real potential of IoT where it can be used by millions of independent devices co-operating together in large or small combinations, and in shared or separated federations. This is not just a networking platform but also a set of commonly agreed methods of information exchange and operation. While the current Internet is a collection of rather uniform devices, however heterogeneous in some capabilities but very similar for what concerns purpose and properties, it is to be expected that the IoT will exhibit a much higher level of heterogeneity, as objects totally different in terms of functionality, technology and application fields will belong to the same communication environment [2].

5.1.2 Emerging Trends

The seamless connection of devices, sensors, objects, rooms, machines, vehicles, etc., through fixed and wireless networks is envisaged to bring in economic value to new applications. Such applications have immediate relevance for transport through intelligent cars, logistics and traffic systems, for environment through smart buildings, for security systems, generating large

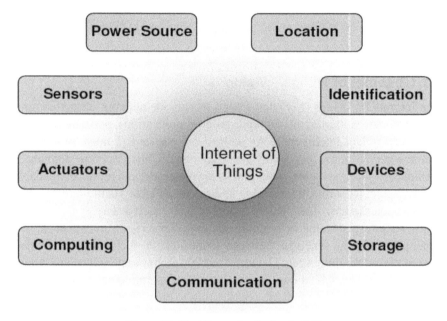

Figure 5.1 Enabling technologies for IoT

efficiency gains in the wider economy. From the technological perspective, the IoT will enable computing to melt invisibly into our business, personal and social environments, supporting our economic, health, community and private lives.

RFID is the most mature of the enabling technologies with established standardization protocols and commercial applications reaching the wider market. The market projections of RFID and related applications are manifold. The global market for RFID products and services is growing fast. However, this is less significant compared to the total revenues expected over the medium to long term, with the spread of smart cards and RFID in all kinds of consumer products, including mobile phones.

Wireless sensor networks as another component of IoT are widely used in industries such as automotive, homeland security, medical, aerospace, home automation, remote monitoring, structural and environmental monitoring. Estimates of their market potential vary, but analysts forecast that falling prices will increase the volume dramatically. Meanwhile, robotics is expanding into new markets. At present, the market share of industrial robotics is larger than that of personal and service robotics, but this is set to change, as the personal robotics segment is expected to lead future market growth. Changing business strategies is the name of the game, in particular in the retail, automotive and telecommunication industries. Firms are embracing the underlying technologies of the Internet of Things to optimize their internal processes, expand their traditional markets and diversify into new businesses.

Location and identity based services are going to be key feature of new Internet services. New opportunities will be created and new needs will be met, bringing about potentially disruptive business models, and new societal services that will improve the quality of life. Future networks will have to deal with an increase in traffic as today's off-line objects are brought online to make industrial processes more efficient with higher degrees of productivity.

Some of the relevant trends in development and deployment of IoT based applications are:

- *Miniaturization of devices*: While VLSI technologies are moving towards fewer microns, Micro Electric Machines (MEMS) and nano electronics are making the sensors smaller and more power efficient so that they can be used inside smaller devices and even living systems.
- *IPV6 as an integration layer*: Drawing from the OSI layer structure and protocols from Internet, it is natural to extend the IP layer as an integration layer for Internet of things as well. But unlike desktops and laptops the nodes attached to the "things" are low in resources such as battery and computing power. However there have been successful attempts to develop a standard named 6LoWPAN, which is an acronym of IPv6 over Low Power Wireless Personal Area Networks, bringing IPV6 over IEEE 802.15.4.
- *Mobile phones as data capturing devices*: Another development that is taking place and beginning to spread globally is to extend data capture systems on mobile phones. Camera, NFC readers and Bluetooth are used for many diverse data capture applications. The advantage of the mobile phone as the pervasive device with support for Internet connectivity will enable its wide application and acceptance. For example, there are mobile applications which use cameras to read 2-D barcodes and LCD screens to display barcodes such that it can be interfaced to registration/payment systems.
- *Low power and energy-neutral devices*: The energy required to operate the intelligent devices will be dramatically decreased. The search is for a zero level of entropy where the device or system will have to harvest its own energy. In addition to relatively established solar cells, there are

success stories such as vibration energy harvesting (perpetuum), powering from RF beams (Powercast), energy from shakes (Sandia national Laboratories) and many attempts by exploiting wind, heat, strain, etc., from energy storage perspective thin film batteries (Cymbet), thin film micro energy cells (Infinite Power Solutions).

- *Handling High Volume and high bandwidth data*: Just to give some numbers, business forecasts indicate that in the year 2015 more than 220 Exabytes of data will be stored. As current network are ill-suited for this exponential traffic growth, there is a need by all the actors to re-think current networking and storage architectures. It will be imperative to find novel ways and mechanisms to find, fetch, and transmit data.

- *Autonomic management*: The ever growing complexity of systems will be unmanageable with the traditional manual approach. It will hamper the creation of new services and applications, unless the systems will show self-properties, such as self-management, self-healing and self-configuration. Towards higher reliability and resilience intelligent systems are being adopted for autonomic management. Distributed intelligence and cooperative approaches are an observed trend in this direction.

5.1.3 Applications

As the technologies needed for the Internet of Things become available, a wide range of applications will be developed. These can support policy in areas including transportation, environment, energy efficiency and health. Huge benefits will come not only from faster productivity growth, but also in many other ways like increasing efficiency in material handling and general logistics, efficiency in warehousing, product tracking, efficiency in data management, reducing production and handling costs, speeding the flow of assets, anti-theft and quicker recovery of stolen items, addressing counterfeiting, reducing mistakes in manufacture, immediate recall of defective products, more efficient recycling and waste management, achieving CO_2 reductions, energy efficiency,, improved security of prescription medicine, and improved food safety and quality.

Major application areas include the following [3, 4]:

- *The Manufacturing, Logistics and the Retail sectors* have already seen a lot of RFID deployments and will continue to be a major user of RFID, wireless sensors and smart objects. Applications include product authentication and anti-counterfeiting, next generation industrial automation and supply chain management, inventory management, Track & Trace, and remote maintenance, service & support.

- *Energy and Utilities sector* includes smart electricity and water transmission grids, real time monitoring of water supply and sewage systems, etc. In addition to the utility infrastructure monitoring, there are many upcoming applications at the consumer end. Efficient energy and water consumption at homes enabled by connected devices to the grid are a major upcoming application area.

- *Intelligent Transportation Systems* applications include support for vehicular ecosystems, use of in-vehicle sensor networks, telematics, GPS and wireless networks for developing smart vehicles and transportation systems. Vehicle to vehicle communications and vehicle to roadside communications can be used for collaborative road safety and efficiency. Vehicle tracking, traffic data collection for management, traffic rule enforcement systems, automotive infotainment systems are going to be a part of an integrated network.

- *Environment monitoring applications* are one of the most talked about applications of sensor networks. This will involve use of wireless sensor nodes for monitoring of weather, environment, civil structures, soil conditions, etc. and will have extensive use in agriculture, weather monitoring, security and surveillance, disaster management, etc.

- *Home management and monitoring* applications will involve use of wireless sensor nodes, smart appliances, wireless networks, home gateways and Internet for applications as wide ranging as home security, elderly care, smart energy control, etc.

There are several other such applications already in use that impact society and economy and it is expected that potentially many more will evolve as various IoT related technologies mature [5, 6].

The development of the Internet of Things, based on the synergetic combination of several scientific disciplines and technologies, creates tremendous opportunities for improving economic competitiveness and citizens quality of life. IoT also raises complex non-technical issues, especially with respect to ethics, privacy, security, governance, spectrum, interoperability, and more, which will be discussed later in this chapter.

5.2 Architectural Framework

The task of seamless integration of 'things' to the Internet will be challenging and a homogeneous approach to it may not be feasible. Major factors of influence are the connectivity, power sources, form factor, security, geographical factors and cost of deployment and operation. Applications with different constraints on these factors will have different optimum architectures for integration. Also the interests of major proponents of specific standards and devices play a major role in creating an ecosystem for specific approaches.

Connecting with the physical world involves the interfacing of everyday objects with various data capturing means. They include mainly identity capture methods such as barcodes, RFID, biometrics, and sensors for physical features such as audio, vision, temperature, pressure, humidity, light and so on. Connecting them to the Internet will involve the integration of multiple connectivity options based on the constraints described above. In general it is envisaged that the integration will be in a hierarchical manner where sensor clusters at the lowest level connect to a suitable access network to reach the Internet. These lowest level networks are termed as edge networks or capillary networks. Capillary Networks is a term introduced by France Telecom to capture the concept of short-distance, edge networks extending existing networks and services to all devices equipped with sensors and actuators and the physical environment in general [7].

One of the major trends is to bring the Internet protocols as close to the end devices/objects as possible with appropriate customization and extensions to the protocols and standards involved. This will enable a unified connectivity of objects to the fabric of Internet as depicted at a high level in Figure 5.2.

5.2.1 Layered Architecture for IoT

One of the major aspects for success of the Internet is attributed to its layered architecture. Layered architecture provides a structured and modular approach for building up the integration framework with multiple compatible components. The high level architecture described above could be mapped into a set of well defined layers as shown in Figure 5.3.

Edge layer consists of technologies involved in interfacing with the end devices and their neighborhood. This will include sensor technologies, wireless communication and networking protocols such as Body Area Networks (BAN), Personal Area Networks (PAN), and Local Area Networks (LAN).

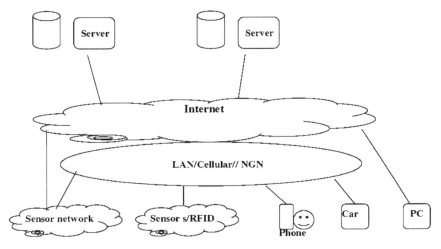

Figure 5.2 High level vision of IoT

RFID, NFC, ZigBee, Bluetooth, UWB, and Wi-Fi are some of the prominent communication standards considered here.

The access gateway layer consists of a collection of network gateways providing connectivity between edge layer and Internet layer. The network devices of a layer will support the Internet protocols and extend it further down to the access network at the edge layer as applicable. They may have the roles of routers, cluster heads, sink nodes, etc., as the underlying edge network support provided. Based on the geographical distribution the networking at this layer may involve Metropolitan area networks (MAN) Wide Area Networks (WAN), Cellular networks, Satellite communication networks, etc., for interconnecting the gateway nodes.

The Internet layer will provide and extend the support of familiar Internet protocols for networking and management. The notion Internet of Things and services encompasses a wider range of devices that use TCP/IP and protocols on top of that to communicate with PCs or other devices. A lot of these devices can use web service technology as a means to communicate or synchronize themselves with other 'things' on the Internet. Web Services can be used to wrap part of the devices' functionality in order to allow bidirectional communication between back end ERP services of the organization/enterprise. Support of IPV6 to enable the scalability and management is a major drive here.

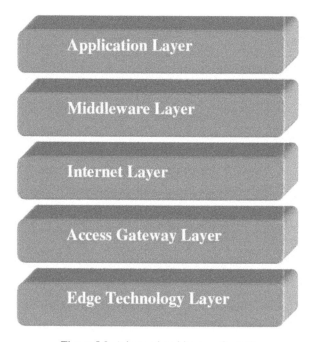

Figure 5.3 A layered architecture for IoT

The middleware layer provides standardized application programming interfaces for abstracting data connectivity, network configuration & management, security, directory services, events handling, etc. The objective of middleware is also to abstract the differences in data formats, addressing and configuration of underlying layers. Some of the autonomic networking features will be driven from here.

A variety of applications requiring the data access and visualizations are taken care of at the application layer through standard application development frameworks, and libraries.

Some of the lessons and experiences from RFID applications could form a basis for the architectural exploration for IoT. EPC Global and ISO are the major active standardization bodies active in the area of RFID [8].

A detailed discussion on the EPC network architecture and its components are given in a later section.

5.2.2 IoT Architecture Trends

In order to make the IoT reasonably cost-effective and simple to deploy for almost every kind of object, we need to agree on a set of Machine-to-Machine (M2M) interface standards for the complete communications package. A unified architecture integrating RFID and ubiquitous sensor networks is to be attempted here.

The standards need to address:

- Web Services as a common platform to publish service definitions and exchange configuration information between nodes.
- Messaging Services Layers providing a basic web-services messaging framework between hosts which abstracts lower layers.
- Common Data Exchange Formats for the sharing of structured data across different systems.
- Internet Protocol Layers or an IP proxy layer, which is used to connect network nodes across multiple types of networking technologies, RF waveforms, and radio platforms.

An architectural framework of IoT needs to incorporate all the desired aspects of an IoT system in terms of scalability, flexibility, adaptivity and future-proof. It needs to specify the standards, components and interfaces for various building blocks such as device interfaces, data formats, networking standards and protocols, service platforms and application interfaces.

Figure 5.4 shows a high level architecture of the IoT vision as promoted by a European consortium initiative CASAGRAS. At the lower level there are clusters of devices hosting sensors, RFIDs, etc., networked together (edge network) and are connected to a wide area network through respective gateways. The back end connectivity could be wired or wireless. Above that there is a middleware layer to abstract the underlying heterogeneity and provide a unified service interface to an application layer supporting multiple diverse applications.

There are well developed architectural frameworks which can contribute towards addressing some of the issues faced by IoT. The architecture from EPC Global for RFID systems is a good reference point to start the buildup of IoT. A detailed discussion on this is presented in later sections.

5.3 Issues and Challenges

IoT needs to be treated as an evolutionary technology and innovative research is required to realize its vision. Contrary to the perception of many, it is not

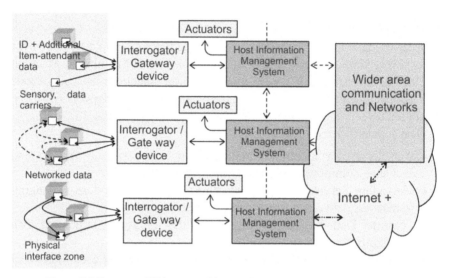

Figure 5.4 Internet of Things – architectural components (adapted from [27])

just a simple reengineering of existing technology. Trillions of connected devices can push current communication technologies, networks and services approaches to their limits and require new technological investigations. In particular, research is required in the field of system architectures, protocols, device technologies, service oriented architecture supporting dynamic environments and security and privacy. Research is also required in the field of dedicated applications integrating these technologies within a complete end-to-end system.

Many of the opportunities and challenges discussed in earlier sections pose specific problems for the research community to solve. Major aspects of research can be categorized into the following:

- Architectural framework for IoT.
- System integration & management.
- Ubiquitous networking.
- Security & privacy.
- Devices and platforms.
- Impact of IoT on environment.

Some of the aspects of these challenges are discussed in detail in subsequent sections.

5.4 Ubiquitous Networking

Over the past decade, there has been an increasing trend towards integrating sensing, communication, and computation into the physical world. No longer restricted to the office desktop, computers are becoming embedded in all aspects of our everyday lives, varying from electronic toys to smart cars, from augmented classrooms to intelligent homes. These computers are also becoming increasingly aware of the environments and situations in which they are used, using factors as simple as the current humidity and light level to as complex as who is using the computer and where it is being used. This push towards ubiquitous computing offers tremendous gains in coordination, safety, and efficiency in domains as diverse as real-time monitoring of soil conditions [9], helping patients with Alzheimer's disease [10], and support for emergency responders [11].

Ubiquitous networking is an integral aspect of networked "things" to provide any time anywhere connectivity especially under mobility. Ubiquitous networking should provide robust connectivity in varying geographical configurations, environmental conditions and application requirements.

5.4.1 Communication Standards

There are numerous standards for communications and networking being used and developed worldwide. They can be generally differentiated in terms of data rate, coverage, frequency of operation, local regulatory status, cost of devices, power consumption, popularity, communication security and other features. In addition to this there are a variety of standards for data and media exchange over such networks.

Some of the relevant communication standards for edge networking are summarized below.

Standards for Edge Networking

Depending on the nature of application in terms of scale, data rate, geographical coverage, availability of infrastructure, etc., there can be a set of applicable wireless communication standards. Most prominent standards in the general for edge networking are ZigBee, Bluetooth, UWB, Wi-Fi. Some prominent proprietary systems are ANT and ZWAVE who target extremely low power communications.

In general, the Bluetooth, UWB, and ZigBee are intended for WPAN communication (about 10 m), while Wi-Fi is oriented to WLAN (about 100 m). However, ZigBee can also reach 100 m in some settings. The FCC power spectral density emission limit for UWB emitters operating in the UWB band is –41.3 dBm/Mhz. This is the same limit that applies to unintentional emitters in the UWB band, the so called Part 15 limit. The nominal transmission power is 0 dBm for both Bluetooth and ZigBee, and 20 dBm for Wi-Fi.

Bluetooth, ZigBee and Wi-Fi protocols have spread spectrum techniques in the 2.4 GHz band, which is unlicensed in most countries and known as the Industrial Scientific and Medical (ISM) band. Bluetooth uses frequency hopping (FHSS) with 79 channels and 1 MHz bandwidth, while ZigBee uses direct sequence spread spectrum (DSSS) with 16 channels and 2 MHz bandwidth. Wi-Fi uses DSSS (802.11), complementary code keying (CCK, 802.11b), or OFDM modulation (802.11a/g) with 14 RF channels (11 available in US, 13 in Europe, and just 1 in Japan) and 22 MHz bandwidth. UWB uses the 3.1–10.6 GHz, with an unapproved and jammed 802.15.3a standard, of which two spreading techniques, DSUWB and MB-OFDM, are available.

Since Bluetooth, ZigBee and Wi-Fi use the 2.4 GHz band, the coexistence issue must be dealt with. Basically, Bluetooth and UWB provide adaptive frequency hopping to avoid channel collision, while ZigBee and Wi-Fi use dynamic frequency selection and transmission power control. IEEE 802.15.2 discussed the interference problem of Bluetooth and Wi-Fi. Quantitative measurements of the coexistence issue for ZigBee, Bluetooth, Wi-Fi, and microwave ovens are discussed in [12]. Moreover, Neelakanta and Dighe [13] present a performance evaluation of Bluetooth and ZigBee collocated on an industrial floor for robust factory wireless communications.

Two connectivity topologies are defined in Bluetooth: the piconet and scatternet. A piconet is a WPAN formed by a Bluetooth device serving as a master in the piconet and one or more Bluetooth devices serving as slaves. A frequency-hopping channel based on the address of the master defines each piconet. All devices participating in communications in a given piconet are synchronized using the clock of the master. Slaves communicate only with their master in a point-to-point fashion under the control of the master. The master's transmissions may be either point-to-point or point-tomultipoint. Also, besides in an active mode, a slave device can be in the parked or standby modes so as to reduce power consumptions. A scatternet is a collection of operational Bluetooth piconets overlapping in time and space. Two piconets can be connected to form a scatternet. A Bluetooth device may participate in several piconets at the same time, thus allowing for the possibility that information could flow beyond the coverage area of the single piconet. A device in a scatternet could be a slave in several piconets, but master in only one of them.

ZigBee provides self-organized, multi-hop, and reliable mesh networking with long battery lifetime [14–16]. Two different device types can participate in an LR-WPAN network: a full-function device (FFD) and a reduced-function device (RFD). The FFD can operate in three modes serving as a PAN coordinator, a coordinator, or a device. An FFD can talk to RFDs or other FFDs, while an RFD can talk only to an FFD. An RFD is intended for applications that are extremely simple, such as a light switch or a passive infrared sensor. They do not have the need to send large amounts of data and may only associate with a single FFD at a time. Consequently, the RFD can be implemented using minimal resources and memory capacity. After an FFD is activated for the first time, it may establish its own network and become the PAN coordinator. All star networks operate independently from all other star networks currently in operation. This is achieved by choosing a PAN identifier, which is not currently used by any other network within the radio sphere of influence. Once the PAN identifier is chosen, the PAN coordinator can allow other devices to join its network. An RFD may connect to a cluster tree network as a leave node at the end of a branch, because it may only associate with one FFD at a time. Any of the FFDs may act as a coordinator and provide synchronization services to other devices or other coordinators. Only one of these coordinators can be the overall PAN coordinator, which may have greater computational resources than any other device in the PAN.

ZigBee networks are primarily intended for low duty cycle sensor networks ($<1\%$). A new network node may be recognized and associated in about 30 ms. Waking up a sleeping node takes about 15 ms, as does accessing a channel and transmitting data. ZigBee applications benefit from the ability to quickly attach information, detach, and go to deep sleep, which results in low power consumption and extended battery life.

The IEEE 802.11 architecture consists of several components that interact to provide a wireless LAN that supports station mobility transparently to upper layers. The basic cell of an IEEE 802.11 LAN is called a basic service set (BSS), which is a set of mobile or fixed stations. If a station moves out of its BSS, it can no longer directly communicate with other members of the BSS. Based on the BSS, IEEE 802.11 employs the independent basic service set (IBSS) and extended service set (ESS). The 802.11 family of standards include 802.11b/a/g and 802.11n giving progressive data rates from 11Mbps to 100 Mbps.

All the above mentioned protocols have encryption and authentication mechanisms. Bluetooth uses the E0 stream cipher and shared secret with 16-bit cyclic redundancy check (CRC), while UWB and ZigBee adopt the advanced encryption standard (AES) block cipher with counter mode (CTR) and cipher block chaining message authentication code (CBC-MAC), also known as CTR with CBCMAC (CCM), with 32-bit and 16-bit CRC, respectively. In 802.11, Wi-Fi uses the RC4 stream cipher for encryption and the CRC-32 checksum for integrity. However, several serious weaknesses were identified by cryptoanalysts, any wired equivalent privacy (WEP) key can be cracked with readily available software in two minutes or less, and thus WEP was superseded by Wi-Fi protected access 2 (WPA2), i.e. IEEE 802.11i standard, of which the AES block cipher and CCM are also employed. Table 5.1 summarizes the main differences among the four protocols where each protocol is based on IEEE standards.

In summary, Bluetooth and ZigBee are suitable for low data rate applications with limited battery power (such as mobile devices and battery-operated sensor networks), due to their low power consumption leading to a long lifetime. On the other hand, for high data rate implementations (such as audio/video surveillance systems), UWB and Wi-Fi would be a better solution because of their data rate support and better efficiency at higher data rates.

Near Field Communications (NFC)

The Near Field Communication (NFC) Forum standards and applications focus on 13.56 MHz RFID technology, whereas other mobile phone standards (being developed by ISO) also addressed UHF technology. These standards use an RFID enabled mobile phone as the data capture device, with a tag based on the contactless smart card ISO/IEC 14443 standard. Currently the identifiers and network infrastructure specified by the NFC Forum differ, for example from those of EPC global, even when EPC data is encoded. Although there is a significant lack of interoperability between NFC Forum and EPCGlobal applications, the risk of a systems clash is relatively low because of the fundamental differences between the way the two systems have been designed. With the potential of NFC-enabled phones being as common as camera phones are today, it is a development that cannot be ignored for its potential to provide services to consumers.

IP Based Convergence

As it is obvious that the framework of Internet of Things has to deal with many communication standards and protocols, the need for a convergence

Table 5.1 Parametric comparison of wireless protocols

Standard	Bluetooth	UWB	ZigBee	Wi-Fi
IEEE spec.	802.15.1	802.15.3a	802.15.4	802.11a/b/g
Frequency band	2.4 GHz	3.1–0.6 GHz	868/915 MHz; 2.4 GHz	2.4 GHz; 5 GHz
Max signal rate	1 Mb/s	110 Mb/s	250 Kb/s	54 Mb/s
Nominal range	10 m	10 m	10–100 m	100 m
Nominal TX power	0–10 dBm	–1.3 dBm/MHz	(-25)–0 dBm	15–20 dBm
Number of RF channels	79	(1-15)	1/10; 16	14 (2.4 GHz)
Channel bandwidth	1 MHz	500 MHz–7.5 GHz	0.3/0.6 MHz; 2 MHz	22 MHz
Modulation type	GFSK	BPSK, QPSK	BPSK (+ ASK), O-QP	BPSK, QPSK, COFDM, CCK, M-QAM
Spreading	FHSS	DS-UWB, MB-OFDM	DSSS	DSSS, CCK, OFDM
Coexistence mechanism	Adaptive freq. hopping	Adaptive freq. hopping	Dynamic freq. selection	Dynamic freq. selection, transmit power control
Basic cell	Piconet	Piconet	Star	BSS
Extension of the basic cell	Scatternet	Peer-to-peer	Cluster tree, Mesh	ESS
Max number of cell nodes	8	8	> 65000	2007
Encryption	E0 stream cipher	AES block cipher (CTR, counter mode)	AES block cipher (CTR, counter mode)	RC4 stream cipher(WEP), AES block cipher
Authentication	Shared secret	CBC-MAC (CCM)	CBC-MAC (ext. of CCM)	WPA2 (802.11i)
Data protection	16-bit CRC	32-bit CRC	16-bit CRC	32-bit CRC

point to abstract out the heterogeneity is highly desirable. The emerging and obvious choice is Internet Protocol (IP), in spite of its various deficiencies with respect to the support of advanced QoS mechanisms.

Major advantages of IP are:

- Extensive interoperability: support for many communication standards. Moreover the next generation wireless technologies envisaged to be an all IP network.
- Established security: It provides a host of proven security protocols for Authentication, access control, and firewall mechanisms.
- Established naming, addressing, translation, lookup, discovery.
- Established proxy architectures for higher-level services including Network Address Translation (NAT), load balancing, caching, mobility, etc.

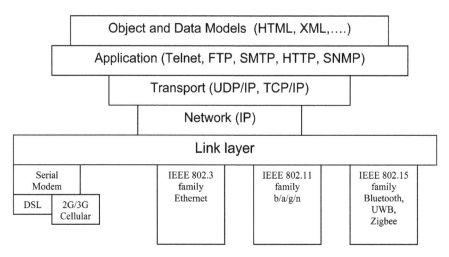

Figure 5.5 IP as an integration layer for multiple edge networking technologies

- Established application level data model and services such as HTTP, HTML, XML, SOAP, REST, Application profiles.
- Established network management protocols and tools such as Ping, Traceroute, SNMP, OpenView, NetManager, Ganglia, etc.
- Transport protocols providing End-to-end reliability in addition to link reliability and QoS mechanisms.

Figure 5.5 shows the integration architecture of these edge technologies to the standard Internet protocol structure. Here IP as the integration layer enables many of the existing applications and standards to continue working in the integrated environment.

Role of IPV6 and 6LoWPAN

The value of the IoT is in its provision of connectivity to "things" for inter-action and information gathering. Drawing parallel with the internet protocol structure there is a push to run the IoT also over IP. IPv6 (Internet Protocol Version 6) is the next generation Internet Layer protocol for packet-switched internetworks and the Internet. The major drive and interest from IoT context for IPV6 is its provision of large address space which can accommodate the individual addressing of all possible objects. The penetration of the IPv6 is in progress all over the world, though at a slower pace than desired [17].

Internet Protocol (IP) has been recognized as a successful integration protocol to be inherited from Internet, in spite of its deficiencies on various

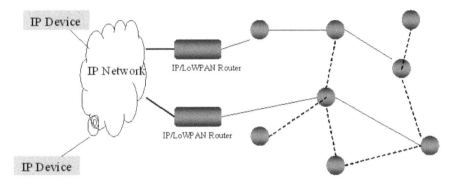

Figure 5.6 LoWPAN extended IP Network

aspects, IoT also can consider adopting it appropriate level for convergence. IPV6 over low power WPAN (6LoWPAN) is an attempt towards this by bringing IPV6 over IEEE 802.15.4 (underlying IEEE standard for ZigBee). This should be extended and evaluated for other edge standards as well.

Cisco, Atmel, the Swedish Institute of Computer Science (SICS) and other leading technology vendors and users formed the IP for Smart Objects (IPSO) Alliance and announced, the availability of uIPv6, one of the world's smallest open-source, IPv6- ready protocol stack, which could enable every device, no matter how limited by power or memory to have an Internet Protocol address, thus promoting the "launch of the Internet of Things" [18], see Figure 5.6.

6LoWPAN

The development for Smart Objects is being undertaken by a group within the IETF [19], with a focus on developing a protocol for IPv6 over Low-Power Wireless Personal Area Networks (6LoWPANs). This enables the transmission of modified IPV6 packets over an IEEE 802.15.4 network. This is a basic protocol that attempts to reduce and compress some of the buffering requirements of IPV6, and to fragment messages. Two RFC documents have been published, one dealing with the basic concept of the 6LoWPAN protocol and the other dealing with ways to reduce the size of messages so that such devices can function in conjunction with the IPV6 protocol.

To make an 802.15.4 subnet under IPV6 several trade-offs are to be considered. The maximum frame size of 802.15.4 is 127 bytes whereas the IPV6 header size itself is 40 bytes. The typical IP packets are large and have to be fragmented to fit in the 802.15.4 frames. Also the multi-hop transmission of the subnet needs to be supported.

To reduce the overhead due to the header, a compressed format is defined. Some of the redundant information available from 802.15.4 header such as source address, destination address, length, etc., are removed from the IP header. The new header is made as an orthogonal stackable format (with optional fields) so that it is included only if required.

There are many challenges and topic for work in the networking aspect of IoT. Some of them are introduced in the following sections.

5.4.2 Managing Heterogeneity

As introduced earlier IoT generally will be an integration of access networks and core networks, similar to cellular networks. Access networks will have a special category called edge networks for last meter connectivity. The requirements of these edge networks are a major differentiating factor in IoT. The communication devices mounted on objects (things) in general have to be extremely low power and low cost. Ideally they should enable the realization of perpetual devices which can sustain themselves. However, there are certain categories of applications which can relax these constraints. With such diverse requirements at the device connectivity level, the edge network will continue to have multiple communication standards. There needs to be a standardization for edge gateways as an integration point to the access and core IoT network. The IP based framework discussed earlier can be a feasible approach and should be extended to cover other edge network standards as well.

5.4.3 System Integration and Management

It has been observed that the scale of the network is an issue for IoT and managing large scale networks is a problem. The traditional management approach is not cost effective for the kind of systems we are talking about for IoT. The IoT requires self-organizing networking to give "deploy and self-operate", mobility, economy and flexibility as well as network resilience. Self-organizing networks must cope with varying requirements for physical and virtual link topologies. Service discovery and reuse by different applications has very high priority. Different applications may require different quality of services and the network will have to organize with suitable topologies.

A unified approach for naming and addressing the nodes are a major aspect for standardization. If one wants to communicate with a node, one must know where it is. In the Internet, a hierarchy of domain name servers (DNS) allows one to do this. The root server is queried, which directs to another server, and so on, until the one that knows the physical address is found. An extension of this system, object name servers (ONS), is expected to serve RFIDs and the Internet of Things. Some of the relevant schemes are discussed in later sections.

5.4.4 New QoS Requirements

IoT will bring in applications with new QoS requirements which are not directly addressed by many of the legacy standards. There are sensors and actuator networks requiring the transmission of low volume data (couple of bytes) with an end-to-end latency of few milliseconds over large distances involving highly resource constrained devices. Technology and standards need to be developed to address such requirements.

5.4.5 Application Specific Standardization

As has been shown there are innumerous communication standards and some of the later standardization attempts were in selecting an appropriate set for suiting the specific application domain. Examples are in vehicular communications, smartGrid, emergency communications, etc. The scope of some of these efforts was just to get a good mix and match with required customizations or extensions. Major impacts of application specificity are at the data model level or profile level.

5.4.6 Radio Spectrum for Wireless Connectivity

The Internet of Things will predominantly use wireless (radio) technologies, at least for the last meter connectivity, and the allocation of radio spectrum is an important issue. In many countries RFID has licensed bands. However most of the IoT systems still rely on the ISM bands. 2.4 GHz band is a universal ISM band being utilized by many devices and communication technologies and as a result highly congested. This will call for a licensed spectrum allocation. But if this is not done uniformly across many countries, then it will increase the cost of devices due to fragmented volumes. In addition to the traditional methods new approaches including cooperative transmissions and spectrum sharing using cognitive radios.

To summarize, ubiquitous networking is the most important constituent making a major difference to the way things interact. Aspects for further research and development work required in the following for improved cost effective connectivity.

- Integration layer for multiple standards to handle heterogeneity.
- Autonomous self-organising networks and adaptivity.
- Network discovery.
- Network survivability and dependability.
- Network scalability and dynamic routing.
- Naming and addressing strategies.
- Programming platforms and application program interfaces (APIs).

A framework for ubiquitous networking addressing all the above mentioned challenges is required for an efficient IoT infrastructure.

5.5 Ubiquitous Identity

In IoT, identity management no longer applies to individuals only, but extends to services, devices, objects, and virtual entities. However, adequate schemes need to understand the difference in nature and role of the entities involved. Identity is used to establish trust in parties, devices, organizations, etc.

5.5.1 Introduction

The goal of ubiquitous networking is eventually to provide capabilities for connecting all objects in a future network. A simple, unobtrusive and cost-effective system of item identification is important to connect objects (e.g., devices and/or machines) to large databases and networks.

Naming and Addressing Using Object Identity

At the application layer each user/object needs to be referred to using an object identity as shown in Figure 5.7. It may be a name or a composite ID with the name and a set of attributes. At service layer this identity should be associated with service IDs which in turn would be mapped to service identifiers like RFID, content ID, telephone number, URI/URL, etc., after proper authentication. Subsequently for the communication purpose this service Id needs to be each service ID should be associated with communication IDs such as session/protocol ID, IP address, MAC address, etc., through a mapping/binding process [5].

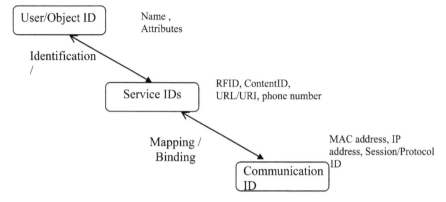

Figure 5.7 Flow of mapping identities at various layers

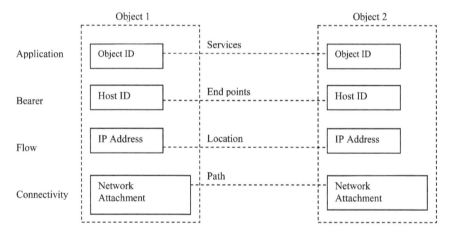

Figure 5.8 Layered mapping architecture between identities

In the scenario of a networked object using IP, the object with an ID at the application layer is attached to a host device having a host ID which resides inside the communication node. This host connects to the network at an attachment point and can use a suitable bearer service for the communication. At application level the connectivity will be referred as bearers and at network level they will be mapped to flows.

The definition of a layered architecture helps in identity mappings/bindings between entities at different levels. An ID resolution server such as Domain Name System (DNS) can provide a function to translate the identifier of object into service/communication ID to access networking services. This process of layered mapping is depicted in Figure 5.8.

The location of an object is an important concept used for identification in many applications. While the identity schemes are independent of location, it is useful to integrate the location information for address binding. Generally, location can be expressed or derived from geographical information such as GPS coordinates or symbolic location such as building numbers or names of neighborhoods. Neighborhoods can also be relative to proximity of objects with known positions. Here neighborhood need not be always physical proximity, but it can be logical similarity as well.

The process of mapping/binding of communication IDs to provide an end to end connectivity is a challenging issue. Also the mapping should enable easy integration of features for service discovery, security and privacy.

Here a few initiatives on mapping IP with other prominent identity standards are illustrated.

IPV6 Binding with EPC Global

One of the prominent identification standards related to RFID is Electronic Product Code (EPC) standard. It has a header identifying the particular EPC version, an EPC manager number indicating the responsible company or organization, an object class number indicating the type of the product followed by a serial number allowing the unique identification of each individual object [18]. The ID length can be 64, 96 or 128.

IPV6 is an addressing scheme for internet devices supporting 128 bit address space and migration from IPV4 is in progress worldwide. When physical objects get integrated to IoT it is natural to explore the use of IPV6 in addressing each individual objects. To support the legacy identity standards like EPC, there can be a mapping between such identity schemes with IPV6 is required and one such approach is discussed in [20].

Mapping of IPV6 with EPC or similar codes have the advantage of the support of standard internet routing protocols while supporting a mechanism for unique identity code. Here the end device could be directly addressable without the need of a network address translator. Also IPV6 has enough address space and scalability is not a problem. From the security point of view also this approach has the benefit of all IPV6 based schemes for encapsulation and authentication.

Host Identity Protocol (HIP) and Object Identification

Host Identity Protocol (HIP) is a development from IETF HIP Working Group and it introduces a new namespace, Host Identity namespace, for Internet. The role of HIP is the separation between the location and identity information by introducing a new cryptographic name space which is called Host Identity (HI). Even though HIP is an end-to-end authentication and key establishment protocol used with IPsec ESP [8], it is designed to support host mobility and end-host multihoming.

Host multihoming and mobility are problems in the current Internet. To support host mobility and multihoming, dynamic readdressing is needed. Additionally, the current IP namespace does not provide identification of hosts, because it is divided into the public and private IP addresses that limit the uniqueness of IP addresses. Furthermore, an IP address is non-cryptographic and thus does not provide any security between hosts. IP addresses do not guarantee data origin confidentiality. HIP introduces a new name space using which all occurrences of IP addresses in applications are eliminated and replaced with cryptographic host identifiers.

5.5.2 Security for IoT

The Internet of Things as an important component of the future Internet will facilitate dynamic, personalized and contextual end-to-end experience which may span multiple trust domains. It will be predominant in a service oriented architecture where future applications need a composition of services in a heterogeneous environment. There will be Internet of Things along with Internet of Services which may be seen as an integration of sensors to a cloud of services. Many early network protocols that are now part of the Internet were designed for performance and not with explicit consideration of security. For example, they lack inherent notions of "identity", "time", and "location" that could contribute to enhancing network security and user accountability and liability. There are many challenges involved in implementing security schemes in such a scale and diversity.

There are two fundamental aspects of security involved namely communication security and other is data security. To meet security requirements the necessary components are confidentiality, Integrity, availability, accountability and liability. Each of the communication standards will have its respective security scheme. An end-to-end communication security need to consider all the participating links and their respective security features.

In many cases data confidentiality is a requirement and it needs restricted access to the information stored on objects. Further, it is required to control whether an object is permitted to participate (i.e. to connect, transmit or receive information) in the network, in general or at any given time. This may be dynamic and context dependent.

Interoperability Requirements for Security

Interoperability is a major issue to be considered in the security of IoT. It has three major aspects namely semantics, grammar and connectivity. The meaning of the data needs to be conveyed for understanding the data exchanged between two objects. Security ontologies and translations mechanisms will be of use here. In addition to this there are several policy languages supporting the grammar of the data exchanged. eXtensible Access Control Markup Language (XACML) is an XML based OASIS Web service oriented standard for communicating access control policies between services. Security Assertion Markup Language (SAML) is a standardised specification for expressing, requesting and delivering assertions regarding the credentials of various entities (users, computers, printers, etc.) under OASIS. SecPAL is a policy and token authoring language developed by Microsoft Corporation combining access control policies and security tokens under the same grammar. The aforementioned grammars could benefit the IoT by providing a standard access control policy language replacing dozens of application-specific languages.

The provision of security in ad hoc network nodes is challenging and probably a third party infrastructure could be envisaged to take care of coordinating the security. The infrastructure can exist as a set of managed components, in order to expose an object or a resource on a network in a dependable way.

5.6 RFID and Internet of Things

RFID is a technology for identifying people and assets without human intervention, enabling computer systems not only to identify objects, but also understand their status. Like smart cards, RFID has a vital role in IoT and its value proposition makes it clear that RFID will be a growing part of the web of identity that is emerging.

Table 5.2 Major components of the EPCGlobal Network

Discovery	Object naming Service	Directs general requests for authoritative product manufacturer information in the EPCGlobal Network
	EPC Discovery Service	Directs requests for trading partner specific data about EPCs, enabling track and trace
Storage	EPC Information Service	Stores and retrieves serial-number-specific product information
Secure Access	EPC Security Framework	Authenticates users identity on the EPCGlobal network, mainly for access control to various information services

5.6.1 RFID Standardization towards IoT

EPC Global and ISO are the major active standardization bodies active in the area of RFID. CASAGRAS Coordination and Support Action for Global RFID-related Activities and Standardization [21, 22] provides a framework of foundation studies to assist the European Commission and the global community in defining and accommodating international issues and developments concerning RFID and the emerging Internet of Things.

5.6.2 EPCGlobal Network Standard for RFID

The EPCGlobal Network^{TM} is a suite of network services that enable the seamless sharing of Radio Frequency Identification (RFID)-related data. It is a visibility architecture model that allows for the creation of private networks enabling business partners to exchange information and enhance efficiency in the global supply chain. It is based on open, royalty free user driven interface standards for identity capture and exchange.

Major components of the EPCGlobal Network are listed in Table 5.2.

EPC Global Network Architecture

EPC Global is a prominent network service architecture primarily for RFID and it has the potential for extending its architecture to cover the whole IoT scope.

A simplified architecture involving RFID tags that comprises four levels is shown in Figure 5.9. Passive tags, such as Class-1 Generation-2 UHF RFIDs, are standardized by the EPC Global consortium and operate in the 860-960 MHz range. Readers are plugged to a local (computing) system, which read the Electronic Product Code (EPC). A local system offers IP connectivity, and collects information pointed by the EPC thanks to a protocol called Object Naming Service (ONS). EPCIS (EPC Information Services) servers process incoming ONS requests and returns PML (Physical Markup

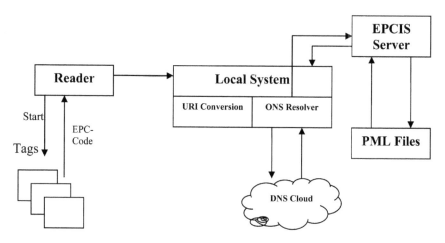

Figure 5.9 EPC Global Network Architecture

Language) files, e.g. XML documents that carry meaningful information linked to tags. The ONS server matches the EPC number, stored on the Tag to the address of the EPCIS server, storing extensive information about the product and tells computer systems where to find information about the object [23, 24].

The Object Naming Service (ONS) is an authoritative directory service that routes requests for information about Electronic Product Codes™ (EPCs) between a requesting party and the product manufacturer. These requests can be routed to a variety of existing, or new, network based information resources. Examples of existing information resources might include Product Information Management (PIM) systems, Enterprise Resource Planning (ERP) Systems, Global Digital Synchronization Network (GDSN) Datapools, or the EPC Discovery Service.

The ONS resolution process takes the Manager ID (unique ID assigned to a manufacturer) and returns one or more network locations (e.g., Uniform Resource Locator or URL) where information about that object resides.

Object Naming Service (ONS)

The Object Name Service can be thought of as a simple lookup service that takes an Electronic Product Code (EPC) as input, and outputs the address of an EPCIS service designated by the EPC Manager in the form of a Uniform Resource Locator (URL). An EPC Manager may actually use ONS to associate several different services, not just an EPCIS service, with an EPC. ONS is architected as an application of the Internet Domain Name System (DNS), which is also a single global lookup service conceptually but is implemented as a hierarchy of lookup services.

ONS works as follows. When an End User application wishes to locate an EPCIS service, it presents a query to its local DNS resolver. The DNS resolver is responsible for carrying out the query procedure, and returning the result to the requesting application. Inside the resolver, a multi-step lookup is performed as follows. First, it consults the Root ONS service controlled by EPCglobal. Then the Root contacts the EPC. The End User then completes the lookup by consulting the Local ONS service, which provides the pointer to the desired EPCIS service.

EPC-IS

The EPC-IS is the network based service that stores, hosts, and provides access to serial number-specific product information that is enabled by RFID. This is one of the many information services to which the ONS may route queries. In some implementations, the EPC-IS provides the full information database that stores this serialized data stream. In other cases, the EPC-IS may only consist of a Web services interface that has been added to an existing product data store. While ONS is only be deployed by manufacturers, EPC-IS may be deployed by all trading partners who wish to track EPCs that move through their field of view.

EPC Discovery Service

The EPC Discovery Service provides track-and-trace capabilities through the EPCglobal Network. It is the registry of every EPC-IS that has information about instances of a certain object or global trade item number (GTIN). As a tagged product moves through the supply chain, it may pass through the fields of view of many different trading partners, each of which may access or record some observable "state" information about that product. Each of these EPC-IS instances then register their "knowledge" with the EPC Discovery Service that services that manufacturer. When track-and-trace information is required for an item, the EPC Discovery Service provides a list of the EPC-IS instances that contain information. Subsequently they can be queried to get the aggregate information on the product in question.

EPC Security Framework

The EPC Security Framework provides a hierarchy of certificate authorities (CAs) and individual trading partner networks. Each application or company in the trading partner community can therefore be issued a digital certificate that can be used at run-time to authenticate their identity. When coupled with the fine-grained access control at the EPC-IS, this allows trading partners to control which of their trading partners have access to what types of information [24, 25].

Sensor Enabled RFID

Here sensors are assumed to be linked to, or attached to, or embedded with the RFID tag and the communication with the application is intended to be through the appropriate RFID air interface protocol. There are two standards groups working on sensors and RFID. IEEE has been responsible for developing various standards for sensors and actuators within the IEEE 1451 series. The latest standard, IEEE 1451.7 specifies the structure and commands for sensors that are used with RFID. ISO/IEC 24753 specifies the rules for configuring sensors and interpreting sensor data. The relevant parts of ISO/IEC 18000 specify the particular air interface protocol rules that apply to a particular type of RFID tag.

The ISO standards have been developed so that item-related data and associated sensor data can be handled in an integrated manner. This means that applications will simply have to specify additional requirements for sensors on top of their basic requirements for encoding data on the RFID tag.

Security Developments and Standards Relating to RFID

Currently there are no formal security standards for RFID and of those available are proprietary. This position is changing, particularly through initiatives taking at ISO to create a specific Work Group within JC1 SC31 to deal with security aspects.

Network and Platform Development Standards for RFID

The EPCGlobal system has a set of standards that define the various interfaces in the network of software and hardware devices between the interrogator and the business application. The standards are supported by implementations, some of which are open source. There are also additional standards that cover aspects of revolving through the ONS to the discovery services. The ISO standards parallel the EPCGlobal network standards, and in fact use them as a core onto which addition features are defined to meet the broad requirements of ISO-related applications. As yet, there are no standards or procedures specified or deployed for resolving ISO based object identifiers, nor for access to discovery services.

A ubiquitous code required for IoT and its associated network supporting different devices with varying application support and standards. A major challenge is in interoperability that such a system can achieve with the EPCglobal and ISO systems.

5.6.3 Major Aspects for New Standards Developments

Major aspects that require attention by standards bodies include all networking layers of the future Internet of Things infrastructure, including the functionality and interfaces for Internet of Things devices, data formats and information codes, naming, addressing and identification issues, middleware aspects and interoperability needs meeting the requirements of the global market place. Interoperability is a particularly crucial dimension in that it contributes to the provisioning of affordable end-to-end solutions while reducing the costs of application integration.

Interoperability is a particularly crucial dimension of standardization that it contributes to the provisioning of affordable end-to-end solutions while reducing the costs of application integration.

The following are the major aspects for new standards developments:

- Ubiquitous identification of Things
- Service Oriented Architecture
    - Directory and Protocols for discovery of services
    - Service description and capability declaration
    - Publication and subscription of devices and resources
- Unified Data exchange formats and protocols
- End-to-End Security
- Middleware with specifications for Application Programming Interfaces
- Convergence layer
- Self-Organisation and Management
- Regulatory framework for RF spectrum and network services
- Data Visualization and User Interfaces
- Resource/Energy management

A systematic development of standards in the above areas will help the stakeholders in developing and deploying IoT applications in a cost effective manner.

5.7 Non-Technical Issues and Challenges for IoT

There are several issues and challenges which are of non-technical in nature facing the operation and management of IoT applications.

5.7.1 Policy Challenges for Privacy and Environmental Hazards

The IoT, through bringing up more and more objects in the networked world, needs mechanisms to respect fundamental rights of citizens and environmental hazards regulated through appropriate policies. Policies should be developed to address the following:

- Make technologies and services affordable, understandable, and accessible to all, in particular to people with special needs or disabilities.
- Ensure that the fundamental rights of individuals to privacy, protection of personal data and consumer protection are adequately captured in the design and functioning of the Internet of Things.
- More electronic materials will become embedded in a range of objects and materials and as an example some materials that are currently either biodegradable or easily recyclable can become problematic.

5.7.2 Governance

Absence of governance may be one major barrier for the widespread adoption of the Internet of Things technology. Without an authority similar to the one that is governing Internet there are high chances that it will be impossible to have a truly global "internet of things". It may not be a major constraint for today, but it will be an issue once the network scales up. Here also the federation model of the Internet may be the choice.

5.8 Conclusions

Vinton Cerf, one of the fathers of the Internet, wrote on the Official Google Blog on 25 September 2008 that "the Internet of the future will be suffused with software, information, data archives, and populated with devices, appliances, and people who are interacting with and through this rich fabric." A few months earlier, he had said in a newspaper interview that billions of Internet-enabled devices with communication capabilities would emerge in the Internet of Things.

In web2: SUMMIT it is discussed that the Web is no longer a collection of static pages of HTML that describe something in the world. Increasingly, the Web is the world – everything and everyone in the world casts an "information shadow", an aura of data which, when captured and processed intelligently, offers extraordinary opportunity and mind-bending implications [26]. Sensor based applications can be designed to get better the more people use them, collecting data that creates a virtuous feedback loop that creates more usage.

It is also observed that industry forums and consortiums have started discussing and developing standards and systems for IoT based applications. This signals that IoT is not a hype but a reality.

The development of the IoT, based on the synergetic combination of several scientific disciplines and technologies, creates tremendous opportunities for improving economic competitiveness and citizens' quality of life.

IoT also raises complex non-technical issues, especially with respect to ethics, privacy, security, governance, spectrum, interoperability, and more, which deserve to catch the highest attention from public authorities, preferably within the context of a sustained and well focused international bodies.

To achieve the vision of communicating things, there are several technical and non-technical challenges to be solved. We have also discussed the impact of IoT for developing countries like India with specific challenges and priorities. Open areas for standardization have been identified where organizations like GISFI could work in conjunction with other global standardization bodies. It is also necessary to bring together the eco-system partners such as government, industry and academics to conduct research and developments to create IPRs and business models to proliferate the IoT technology for mass benefit.

References

[1] Internet of Things in 2020, Roadmap for the future infso d.4 networked enterprise & RFID infso g.2micro & nanosystems in co-operation with the working group RFID of the ETP EPOSS Version 1.1, 27 May 2008.
[2] http://googleblog.blogspot.com/2008/09/next-internet.html.
[3] Disruptive Technologies: Global Trends 2025, SRI Consulting Business Intelligence, Appendix F: The Internet of Things, 2008.
[4] http://ec.europa.eu/information_society/eeurope/i2010/docs/future_internet/swp_internet_things.pdf.
[5] http://inventorspot.com/articles/top_ten_inventions_2008_21971.

[6] http://www.internet2008.fr/spip.php?article9.

[7] http://www.afilias.info/news/2008/10/06/afilias-releases-proposalinternetthings-architecture-advance-eu-meeting.

[8] J. Buckley, From RFID to the Internet of Things. Pervasive networked systems. Report on Conference organized by DG Information Society and Media, Networks and Communication Technologies Directorate, CCAB, Brussels, 6–7 March 2006.

[9] J. Burrell, T. Brooke, and R. Beckwith. Extending ubiquitous computing to vineyards. In *Extended Abstracts on Human Factors in Computing Systems (CHI2003)*, pp. 822–823, 2003.

[10] T. Salvador and K. Anderson, Practical considerations of context for context based systems: An example from an ethnographic case study of a man diagnosed with early onset Alzheimer's disease. In *Ubicomp 2003*, pp. 243–255, 2003.

[11] X. Jiang, et al., Siren: Context-aware computing for firefighting. In *Proceedings of the Second International Conference on Pervasive Computing (Pervasive 2004)*, Vienna, Austria, pp. 87–105, 2004.

[12] A. Sikora and V.F. Groza, Coexistence of IEEE802.15.4 with other systems in the 2.4 GHz-ISM-band. In *Proceedings of IEEE Instrumentation & Measurement Technology Conference*, Ottawa, May, pp. 1786–1791, 2005.

[13] P.S. Neelakanta and H. Dighe, Robust factory wireless communications: A performance appraisal of the Bluetooth and the ZigBee collocated on an industrial floor. In *Proceedings of the IEEE International Confernce Ind. Electron. (IECON'03)*, Roanoke, VA, November, pp. 2381–2386, 2003.

[14] J.S. Lee, Performance evaluation of IEEE 802.15.4 for low-rate wireless personal area networks. *IEEE Trans. Consumer Electron.*, 52(3):742–749, August 2006.

[15] J.S. Lee and Y.C. Huang, ITRI ZBnode: A ZigBee/IEEE 802.15.4 platform for wireless sensor networks. In *Proceedings of IEEE International Conference Systems, Man & Cybernetics*, Taipei, Taiwan, October, pp. 1462–1467, 2006.

[16] J.-S. Lee, Y.-W. Su, and C.-C. Shen, A comparative study of wireless protocols: Bluetooth, UWB, ZigBee, and Wi-Fi. In *Proceedings of the 33rd Annual Conference of the IEEE Industrial Electronics Society (IECON)*, Taipei, Taiwan, 5–8 November, 2007.

[17] TRAI Consultation Paper on Issues Relating to Transition from IPv4 To IPv6 in India.

[18] http://www.ipso-alliance.org/Pages/Main.php.

[19] The Internet Engineering Task Force (IETF), http://www.ietf.org/.

[20] S. Poslad, *Ubiquitous Computing – Smart Devices, Environment and Interactions*, Wiley, 2009.

[21] Research Needs and Trends – CERP – Cluster of European RFID Projects, Report, August 2008.

[22] Functional developments, associated standards and the emerging Internet of Things, WP5 White Paper, CASAGRAS.

[23] EPCGlobal, http://www.EPCGlobalinc.org/home/.

[24] The EPCGlobal architecture framework, http://www.EPCGlobalinc.org/standards/architecture/architecture_1_3-framework-20090319.pdf.

[25] GS1 EPCGlobal Standards for item-level tagging, What's in it for me? http://www.gs1.org/docs/epc/GS1_EPCGlobal_Item_Level_Standards.pdf.

[26] T. O'Reilly and J. Battelle, Web squared: Web 2.0 Five Years On.

[27] A. Furness, CASAGRAS and the Internet of Things. European Centre of Excellence for AIDC.

Biography

P. Balamuralidhar obtained his Master of Technology (M.Tech) in Electrical Engineering from the Indian Institute of Technology, Kanpur in 1987. He has over 20 years of research and development experience in signal processing, embedded systems and wireless communications. He has published over 40 contributions in various international journals and conferences. He has led many R&D programs in wireless communications including reference designs development for 3G-WCDMA and WLAN. Balamuralidhar was the leader of TCS participation in two EU FP6 research consortium programs, namely, My Adaptive Global NET (MAGNET) and End to End Reconfigurability (E2R) in the area of next generation wireless communications. He also leads TCS participation in the Broadband Wireless Consortium India (BWCI) and the collaboration program with the Centre of Excellence in Wireless Technologies (CEWiT). His areas of research interest are in Networked Embedded Systems, Wireless Sensor Networks, and Cognitive Radios & Networks.

Dr. Balamuralidhar is currently Head of the TCS Innovation Lab at Tata Consultancy Services Ltd. (TCS), Bangalore. The major focus of his work is in applications of sensor networks & telemetry in multiple domains including healthcare, utilities, retail, automobiles and buildings. Before TCS he worked with the Society for Applied Microwave Electronics Engineering & Research (SAMEER) and Sasken Communications Ltd.

6

Green ICT

Prateep Misra

TCS, Kolkata, India

It is our responsibility to estimate the GHG emissions from the Information and Communications Technology (ICT) industries and to develop opportunities for ICT to contribute to a more efficient economy.

Global eSustainability Initiative Chair, Luis Neves

Climate change is now the greatest challenge facing the global community, having serious implications on the future of world economy, health & safety, food, security and many other dimensions. Rising levels of greenhouse gases such as carbon dioxide and methane have led to serious and irreversible changes in the earth's ecosystems. It is generally accepted that increasing levels of greenhouse gases (GHG) in the earth's atmosphere are causing a continual rise in the earth's average surface temperature – by some estimates, average temperatures can increase by almost 5°C by the end of the 21st century as compared to the year 2000. Any rise in temperatures above 2°C can cause irreparable damage to Planet Earth and humanity. The ICT community is actively participating in the global scale efforts now underway in meeting this challenge. The ICT community has recognized the need for continuous improvements, new technologies, and disruptive innovations in achieving the climate change related targets in front of the global community. The subject of this chapter, i.e. Green ICT, is the ICT community's contribution towards this momentous task. Indeed, ICT is essential to make it economically possible to achieve necessary reduction in GHG concentration levels in the earth's atmosphere.

R. Prasad (ed.), Future Trends and Challenges for ICT Standardization, 229–269.
© 2010 *River Publishers. All rights reserved.*

We first start with an explanation of what we mean by green ICT. In this chapter we refer to green ICT as a *set of evolving tools, methods, techniques and applications* of Information and Communication Technology that can be used for reducing the carbon footprint of all human activity including the ICT sector itself.

While there are GHG gases other than CO_2 that impact climate changes, we have kept our focus on the impact of the ICT sector on CO_2 emissions caused by use of fossil fuels only. Also, we have kept other important environmental issues such as use of harmful substances and hazardous electronic wastes outside the purview of this discussion. The rest of the chapter is organized as follows.

Section 6.1 describes the carbon footprint of the ICT sector and major contributors of emission from the ICT sector. Section 6.2 focuses on generic approaches for green ICT for reducing energy consumption in ICT infrastructure. Section 6.3 discusses the need of new data centers to cater the increasing demands on space and power requirements. It also focuses on typical electrical components in data centers. Section 6.4 emphasizes green telecom infrastructure and devices. Section 6.5 explains the impact of PCs and peripherals, it also discusses the components of PCs that deserve special attention with respect to energy efficiency – namely displays, power supplies and disks and how intelligent power management techniques can be used to reduce PC power consumption in office environments. Section 6.6 discusses important organizations and programs active in this area in order to promote green ICT and leverage the power of ICT for creating the low carbon economy and society.

Section 6.7 lists some of the key areas where ICT standardization will impact climate change mitigation efforts and how GISFI can contribute. In Section 6.8 conclusions from this chapter are given.

To put it succinctly, ICT has the following effect on carbon emissions:

- the ICT sector emits CO_2, and
- the use of ICT enables the reduction of CO_2 emission in other areas of human activity.

In this chapter we will take a deeper look at both these aspects. With respect to the first, we will drill down into the major components of the ICT sector and their contribution to emissions. We will also look at specific green technologies being adopted in these areas and how they help reduce energy consumption and hence emissions. This chapter will also briefly discuss the various ways in which applications of ICT will enable reduction of carbon footprint of other sectors. The role of standardization in the area of green ICT will also be discussed.

6.1 Carbon Footprint of the ICT Sector

The ICT sector has grown exponentially in the last two decades, transforming society and economy. ICT impacts business, lifestyle and family relationships like never before. As the ICT sector grows, the carbon footprint of the ICT sector will continue to grow. Here are some examples of the explosive growth of ICT:

- It is estimated that there will be more than 3 billion computers connected to the Internet by 2011 [1]. There are predictions that by 2020, the number of devices connected to the Internet will be around 50 billion, forming a part of the so called Internet of Things. Today there are more than 1.5 billion users of Internet. This number is expected to increase significantly as more and more users from developing nations start using Internet, many of whom will access Internet via their mobile phones.
- The number of mobile phone users in India in June 2009 was above 470 million with an annual growth of almost 50% [2].
- For most economies, the share of GDP attributable to the ICT sector is already quite significant and is increasing each year. In India, ICT sector contributed about 5.8% of the national GDP in FY2009. Share of GDP attributable to ICT sector in developed economies such as UK is close to 7% [3].

As of 2007, the ICT sector was responsible for about 2% of total carbon emissions at over 0.8 billion tons of CO_2 equivalent [4]. With the kind of growth happening in the ICT sector, total emissions from this sector are estimated to rise to about 1.4 billion tons by 2020. Figure 6.1 shows the Global ICT footprint.

Major contributors of emission from the ICT sector are as follows:

- Telecom Infrastructure and Devices. As of 2007, 37% of all ICT emission is due to telecom infrastructure and devices. This includes emission caused by mobile network infrastructures, mobile devices, and fixed broadband and narrowband devices. Increasingly as more and more people get access to mobile telephony, the share of emissions from mobile devices and infrastructure will increase and will constitute almost 17% of all ICT emissions.

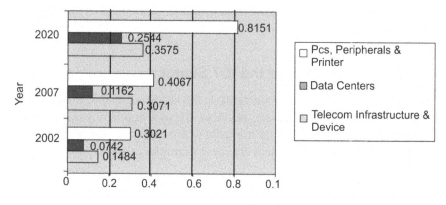

Billion tonnes of Co2 equivalent

Figure 6.1 Global ICT footprints by sector (adapted from [4])

- Data Centers. As of 2007, 14% of all ICT emission is caused by data centers. This includes both corporate data centers and the Internet data centers where large scale consumer web applications such search engines and social networking sites are hosted. Roughly 50% of the emission due to data centers is due to power system losses and cooling loads. The bulk of the remaining part of the emission is caused by the energy consumed to power up low cost commodity servers that now dominate most data centers.
- End User Devices such as PCs and Peripherals. Bulk of all ICT emissions is due to PCs, desktops, laptops and peripherals such as printers. As of 2007, roughly 50% of all ICT emissions are due to PCs and peripherals mainly because of the power consumed in CRT displays. By 2020, it is expected that there will be no CRT displays in use. However, as the number of PCs and laptops are expected to rise to around 4 billion, total emissions from PCs and peripherals will increase in both absolute and percentage terms. By 2020, PCs and peripherals will contribute about 57% of all ICT emissions.

In the sections that follow, we take a deeper look into each of these components.

6.1.1 Role of ICT in Emission Reduction

While climate change issues impose a huge responsibility on the ICT sector to reduce its own carbon footprint intensity and achieve sustainable growth, it also provides unique opportunities for creating an even greater impact by enabling significant emission reduction in other sectors. The Global e-Sustainability Initiative Report [4], estimates that the ICT sectors can enable emission reduction from BAU levels by almost 15% – that is about 5 times the ICT sector's own direct carbon footprint.

Major contributions of the ICT sector in emission reduction would be in the following areas [4]:

- *Smart Power Grids* – The role of ICT in creating Smart Grids cannot be overstated. ICT enables fine grained monitoring of power systems and enables two way communications between end users and power producers. While on one hand it enables power producers and distributors to reduce transmission and distribution losses, it also enables consumers to get a better idea of their own consumption and reduce usage. It is possible to shape the demand of consumers via Demand Response. In addition ICT allows integration of renewable sources of energy and distributed generation.

- *Smart Buildings* – Buildings are responsible for roughly 15% of all emissions. ICT can be used to optimize and monitor the energy consumption right of a building from the design stage to its use. Building automation systems can significantly optimize the use of energy for lighting, heating, ventilation and air-conditioning.

- *Smart Logistics and Intelligent Transportation* – Transportation and logistics account for nearly 14% of GHG emissions, mainly driven by supply chain transportation and storage operations. Supply chain execution systems such as transportation management systems, route planning and optimization, computer aided dispatch, warehouse management systems, etc., together with technologies such as RFID, Global Positioning System and wireless networks help optimize logistics operations with respect to fossil fuel consumption. Also, emission friendly vehicles such as hybrids and Electric Vehicles will increasingly be used over the years. Intelligent Transportation Systems (ITS) will help in improving road safety, increase traffic efficiency, improve freight handling and public transit systems and provide driver assistance. All this will lead to less CO_2 emission from vehicles.

- *Smart Manufacturing & Industrial Operations* – 21% of all GHG emissions are caused by industrial operations. Big contributors are the chemical and petro-chemicals industries, iron and steel plants, cement plants and manufacturing. Process control and automation systems, intelligent motor control & variable speed drives, IT driven energy consumption monitoring, audit & control, production planning and ERP are some of the ways in which ICT brings about system efficiencies and hence energy use and emission reduction.
- *Dematerialization* – This refers to substitution of carbon intensive activities with digital alternatives. Examples include tele-conferencing and video conferencing in lieu of face to face meetings, tele-commuting and "work-from-home" instead of physical presence in place of work, distance education, tele-medicine etc. In addition to this, embodied carbon can be reduced via digitization of media such as paper and CDs, etc.

6.2 Generic Approaches for Green ICT

Having set the context of green ICT in Section 6.1, we now take a look at some generic approaches for reducing energy consumption in ICT infrastructure.

6.2.1 Increasing Visibility

Insight into what is happening in the infrastructure at multiple levels of abstraction and detail is a key requirement. One of the first things that must be put in place is a systematic approach towards energy consumption measurement and a set of performance metrics that capture energy usage efficiency. Measurements and performance metrics are important because

- It enables energy use reduction via change in usage patterns by the consumer. When consumers are aware at a sufficient level of detail, they may adapt, reduce or altogether avoid energy consumption as appropriate.
- It clearly brings out inefficiencies in the system and encourages technology refreshes and optimal system design
- It enables automation in energy saving. Once a measurement system is in place, energy optimization systems may be put in place that automatically sets the system operating points in a way that maximum useful work is obtained for the energy expended.

A good measurement system will have the following characteristics:

- It will provide data at various levels of granularity and abstraction. If the measurement system is able to capture data at the component level, then using techniques such as OLAP it would be possible to provide reports that allow users to do aggregates, drill down and slice/dice energy usage data. Also, by correlating energy usage data with other events, it would be possible to provide customized views of energy consumption for various types of stakeholders. For example, a good measurement system in a data center scenario would provide the following types of reports:

  - Component wise consumption for each server, storage system and switch,
  - Rack level power consumption,
  - Subnet wise power consumption,
  - Application wise power consumption,
  - Customer/Department wise power consumption.

  In case of Telecom Infrastructure, there is a need for making visible how energy consumption is distributed across various systems/sub-systems, components, segments and sub-segments of the network, across various services and applications.

- A good measurement system will balance accuracy, cost and ease of implementation. While it is important to have accuracy, it must be balanced with the cost and ease of implementation. For example in a data center, it is possible to measure energy consumption at the data center, rack and system level using direct methods using energy meters and intelligent power distribution units. It is also possible to estimate power consumption using indirect methods using host utilization metrics such as CPU and memory utilization data provided by the operating system. The latter is less accurate but is easy to implement. A good measurement system will also leverage the instrumentation already in place.

- A good measurement system will be fit to purpose. The measurement system should capture key performance indicators that are appropriate for applications and services in context and the design goals. There are multiple metrics possible for energy usage efficiency. Since the end goal of any energy efficient infrastructure would be to maximize useful work for energy expended, the choice of metrics in a given scenario must be driven by the context in which the infrastructure is used, the applications and customers supported.

6.2.2 Consolidation

System level inefficiencies must be eliminated in order to achieve green ICT. Systems must be operated such that utilization is maximized while maintaining desired service levels. Consolidation is a proven method that increases utilization of infrastructure and helps reduce energy consumption substantially. Put simply, consolidation involves deploying multiple applications/users/services on a common shared virtualized resource pool. Consolidation improves energy efficiency since it enables infrastructure to be used at higher levels of utilization. It also reduces space usage and the embodied carbon footprint.

The problem of low utilization levels is especially prevalent in typical enterprise IT environments. The availability of low cost commodity servers often leads to uncontrolled server sprawl in the data center. It is easy to procure new equipment since the cost of acquisition is low. The downside of this easy availability is that most machines run at low levels of CPU utilization. The overall environment is inefficient in terms of energy and space usage. Examples of consolidation in the data center are as follows:

- Eliminating many smaller machines by replacing them with a single powerful server.
- Running server virtualization software and running multiple Operating System instances on a single hardware unit.
- Running multiple applications on the same server by logical portioning of the server operating system.
- Serving multiple customers on the same application by designing multi-tenant applications .
- Consolidating multiple data centers by having a fewer large sized data centers, each optimized with respect to energy use.

Similarly, network infrastructure may be consolidated via the sharing of resources across multiple service providers – for example sharing of towers, switching equipment, radio equipment and antennas. In fact many countries have mandated sharing of infrastructure by law. Another example of consolidation is via the use of flexible base stations that allow multiple services to be run on the same hardware, such as both GSM/EDGE and WCDMA/HSPA.

6.2.3 Energy Proportional Design

Green ICT requires creation of ICT equipment that exhibit Energy Proportional behavior. What this means is that ICT equipment must consume energy in proportion to the amount of work performed. In an ideal scenario, no power should be consumed when there is no load and the equipment is idle. As the system is loaded, the power consumption should increase in proportion. However in real life this is almost always not the case. A typical server for example consumes almost 50% of the maximum rated power consumption even when it is running at 0% utilization. This behavior of servers is primarily due to non-proportional energy consumption of its components such as memory, disks, network interfaces and power supplies. Some of these components exhibit very low dynamic range of peak power consumption versus idle power consumption [5].

The problem of non-proportional behavior exists in other data center equipment such as switching, storage, power supplies, UPS, air conditioning, air handling systems as well. Non-proportional behavior is also exhibited by PCs, laptops and home networking equipment. The data center as a whole also exhibits non-proportional behavior and total energy consumption in the data center exhibits low dynamic range. Therefore at almost all operating levels other than peak loads, energy efficiency is less than ideal.

6.3 Green Data Centers

Data centers are the fastest growing segment of ICT as far as carbon emission is concerned. Rapid growth in use of Internet, web applications, online services, VOIP, IPTV and enterprise IT needs has resulted in proliferation of data centers. In fact in most cases there is a severe crunch of data center space. Existing data centers are not able to cope with the increasing demands on space and power requirements and new data centers are coming up each day. Figure 6.2 shows typical electrical components in data centers. Web services providers are building cavernous warehouse scaled data centers to meet their growing needs.

In an average web applications data center, power consumption costs together with the burdened cost of power and cooling infrastructure add up to about 40% of total monthly operating costs. About 60 to 70% of the power goes not into the IT equipment, but into facilities and cooling. Cost of a powering and cooling a low cost commodity server over its lifetime often exceeds the acquisition cost. In this section we look at how power is utilized in a data center and some of the best practices that may be adopted for achieving energy efficiency.

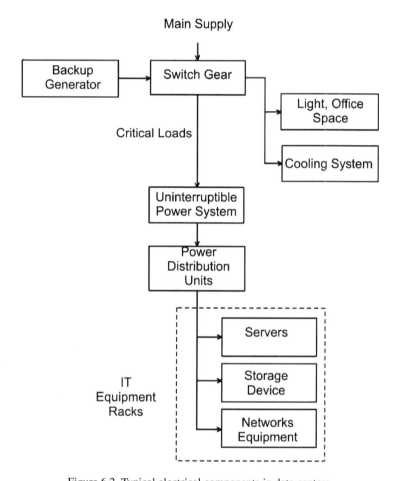

Figure 6.2 Typical electrical components in data centers

6.3.1 Measuring Energy Efficiency

It is important to establish appropriate metrics for evaluating data center energy efficiency. Energy efficiency metrics serve the following purposes:

- It allows benchmarking of data center design and performance with respect to energy efficiency.
- It indicates the efficiency levels in design and operation and helps find scope for improvement.
- It helps taking a decision whether a new data center must be built or not.

Realizing the importance of metrics, The Green Grid – an association of IT professionals – has recommended a set of metrics for energy efficiency. We will discuss two important metrics – Power Usage Effectiveness/Data Center infrastructure Efficiency (PUE/DCiE) and Data Center Productivity (DCP) [6].

Power Usage Effectiveness

Power Usage Effectiveness is defined as the ratio of Total Facilities Power to IT Equipment Power as shown in Figure 6.3, i.e.

$$PUE = \text{Total Facilities Power/IT Equipment Power},$$

Total Facilities Power is the power measured by the utility meter for the data center premises, and IT Equipment Power is the load associated with all IT equipment in the data center including servers, storage, network and displays.

Total Facilities power includes power required for cooling, air handling, power delivery components, UPS, distribution losses and lighting.

Data Center infrastructure Efficiency (DCiE) is the reciprocal of PUE.

PUE/DCiE metrics helps in understanding energy allocation in the data center and gives an indication of efficiency of operation with respect to cooling & air handling energy consumption and power distribution losses. A 100% efficient data center will have a PUE of 1. Most data centers operate between 1.3 and 3.0. Some of the best ones operate around 1.1.

Data Center Productivity

Data center productivity is an evolution of PUE/DCiE in that it measures energy efficiency in terms of useful work performed by the data center. PUE deals exclusively with power consumption and does not give any indication of actual useful work performed by the data center. A data center can be thought of as a black box that gets power as input, generates heat and performs useful work in the process. DCP is an attempt to measure the useful work.

DCP is the ratio of useful work performed to total facilities power. Green Grid has also recommended some measures that measure productivity in terms of total energy consumed in the data center for a given amount of useful work. What constitutes useful work would depend on the context. An example of useful work could be number of the transactions processed by a given application per Joule of energy or Watt of power expended.

6.3.2 Cooling Systems

Cooling systems constitute a significant portion of the total power consumed in the data center and therefore deserve close attention with respect to energy efficiency [7]. Out of the total electrical energy that goes into data center IT equipment, an equivalent amount is output as heat and must be removed quickly so that the data center operates within the operational temperature and humidity envelope specified by the equipment vendor. In most data centers almost an equivalent amount of energy is spent in cooling and associated power distribution equipment.

We now look into specific areas that require attention and corresponding best practices.

Improving Airflow Management

A typical data center consists of a raised floor on which racks are kept in an alternating hot aisle and cold aisle arrangement. Cold air is forced from the bottom plenum through perforated tiles in the cold aisle by computer room air conditioning/handling (CRAC/H) units. The cold air is drawn by the server/device fan and pushed out into the hot aisle.

The hot air in the hot aisle is then removed via overhead openings in the false ceiling into a return air plenum.

The efficiency of cooling depends on how tightly the hot air and cold air flow is maintained and the temperature difference between the hot and cold air. Following are some of the inefficiencies that are common in air handling operations:

- Short circuiting of hot air from the hot aisle to the cold aisle from the top of the server rack.
- Short circuiting of cool air back to air conditioning units bypassing the server racks.
- Short circuiting of hot air to the adjacent cold aisle through empty slots in the racks.
- Poor airflow through IT equipment.
- In adequate pressurization of in the raised floor that does not cause proper airflow.

By ensuring proper air handling it is possible to reduce power consumption related to cooling. Proper design of the hot and cold aisle with proper physical separation of the hot and cold areas prevents mixing of hot and cold air. Sealing leakages in the under-floor plenum, blanking of unused space in racks, clearing all obstruction to airflow and efficient collection of hot air

PUE: Power Usage Effectiveness
DCE: Data Center Efficiency

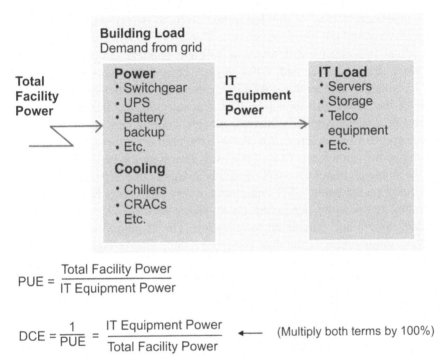

$$PUE = \frac{\text{Total Facility Power}}{\text{IT Equipment Power}}$$

$$DCE = \frac{1}{PUE} = \frac{\text{IT Equipment Power}}{\text{Total Facility Power}} \quad \longleftarrow \quad \text{(Multiply both terms by 100\%)}$$

Figure 6.3 Illustration of how PUE and DCE would be calculated in a data center (adapted from [6])

in overhead plenums is necessary. It is also required to carefully plan the position of perforated tiles and placement of CRAC units. Often it may be necessary to use computational fluid dynamics modeling to optimally place the CRAC units and perforated tiles.

Data Center Temperature

It is possible to lower energy costs by simply running the data center at a higher temperature level. Most IT equipment has a certain operating envelope and it is not necessary to maintain the data center ambient temperatures below/outside this envelope. Thus increasing the average temperature setting of the data center will result in lower cost of energy for cooling. Most data centers are run at a temperature setting between 20 and 23°C. It has been suggested that we may easily set it up as high as 26°C and thereby save on cooling costs [11].

However, recent research [8] has shown that it is necessary to look into the temperature setting of the data center from an overall energy efficiency point of view and not just cooling costs. Certain IT equipment such as CPUs and server fans behave differently at different ambient temperatures as far as power draw is concerned. CPU leakage current increases with increase in CPU junction temperature. As CPU leakage current increases, the power draw of the server also increases. The heat rejected into the room also increases as a result. Similarly, in some servers, cooling fans begin to move at higher speeds to compensate for increase the ambient temperature increases to achieve the same levels of heat removal. Therefore the benefits of increasing the ambient temperature setting may be marginal or even negative at times depending on the nature of IT equipment. Every data center has an optimum operating temperature depending on its characteristics, IT equipment, and cooling system and outside air temperature. The technique of using higher ambient temperature in the data center to reduce energy consumption may provide benefits in data centers that are already working at high PUE levels, for example those that use outside air economizers with no chiller based cooling. This is possible in cool climate regions where outside air cooling is possible.

Outside Air Economizers

Outside air economizers are increasingly being used in large data centers to achieve very high PUE values [9]. It involves the use of cold outside ambient air to cool the data center instead of using of chillers and CRAC. The return air from the data center is mixed with the outside ambient air by the Air Handling Unit and the air supplied to the cold aisle of the data center is maintained at a constant temperature. Typically a temperature of about 24°C is maintained in the supplied air. This air is drawn through the server where it removes the heat and in turn gets heated up. The expelled hot air is again drawn in to the AHU where it is again mixed with outside air. This method is also referred to as "Free Cooling".

The use of air side economizers is especially attractive in regions where the climate is cool more or less throughout the year – with no long periods of high temperature. For example, Microsoft has set up its data center in Dublin where the temperature seldom rises above 27°C. The temperature in the aisles may be maintained at temperature that is close to the upper limit of the operating range of temperature and humidity specified by server vendors.

Typically, this would be around 35°C. In most moderate regions of the world, it would therefore be possible to supply the data center with 100% outside air for most days of the year. On the days of extremely high temperatures, chillers can be used to cool the outside air to lower temperatures before passing it on to the server room.

Data centers have improved the PUE by over 50% by use of outside air economizers. Data centers with PUE value less than 1.2 in most cases use Free Cooling. Free Cooling may also be used in Radio Base Stations in mobile networks.

Direct Water Cooling

A promising technology that may be used for reducing data center cooling costs is the use chilled water to cool servers directly [10]. Traditionally chilled water is used to cool the air which in turn is used to cool the server. In direct water cooling, chilled water is delivered to the server rack in a water cooled door that is placed in the back of the rack. An example of this is the IBM rear door heat exchanger racks. This method gives a boost to the cooling efficiency. Most data centers already have chilled water for air conditioning and delivery of cool water closer to the CPU should not pose a big problem. The advantage of direct water cooling would be huge improvement in cooling efficiency, one that has the potential of increasing PUE to very high levels.

Variable Speed Fans and Pumps

Significant energy savings may be achieved by using energy efficient fans and pumps. The torque required to rotate a centrifugal fans or pumps is proportional to the square of the speed. The power required is then proportional to the cube of the speed. The temperature control system adjusts the flow rate or pressure of the chiller water flow or air flow in CRAH as the measured temperature of the server room changes. In absence of variable speed drives, the control is achieved using mechanical valves or dampers that are highly energy inefficient. Control using variable speed drives for CRAH fans and chiller pumps can significantly reduce energy consumption of cooling systems of data centers.

6.3.3 Electrical Distribution

A green data center needs to maximize the power distribution and switching efficiency since power distribution networks contribute a significant component of the energy losses.

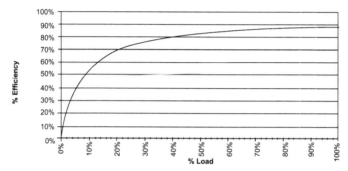

Figure 6.4 UPS efficiency curve (adapted from [13])

Losses in power distribution happen at every point in the power distribution network [12]. There are losses in each of the transformers that step
down the supplied utility HT supply to the 280/240 V levels for use at the
IT equipment. If there are three transformers that are used in this process,
then losses happen at each of the transformers. Additional losses happen in
the UPS and Power Distribution Units (PDUs). The PDU typically has a step
down transformer that adds to the loss. Finally the power supply unit in the
IT equipment itself works at less than 100% efficiency and thus contributes
to the overall losses.

We now estimate the overall losses in electrical distribution at 50% load
conditions. A high efficiency transformer works at around 99.7% efficiency.
Typical high efficiency UPS works at around 95% efficiency. PDU and wiring
efficiency is of the order of 99% in a high efficiency scenario. As far as IT
equipment power supplies are concerned, a figure of 90% may be considered.
Thus the overall efficiency of a system that includes one step down transformer action from HT supply to UPS is something like this: HT Transformer
× UPS × Distribution × IT Power Supply = 0.997 × 0.95 × 0.99 × 0.9 =
0.84. Thus we find that overall efficiency is around 84%.

UPS Efficiency

Choice of UPS and overall design of UPS system has a great impact on data
center efficiency [13]. The choice of UPS must be governed by the efficiency
of the UPS at realistic loads as shown in Figure 6.4. Efficiency metrics quoted
at full load are not relevant since in no case is the UPS run at full load.

Any UPS has three components of losses – losses that are independent of
load, losses that are proportional to the load and losses that vary as the square
of the load. When a UPS system is set up in parallel redundant configuration

such as $N + 1$ and $2N$ configurations, then the load on each UPS is less by an amount dependent on the redundancy level. In case of $2N$ configurations, 50% is the maximum load on each UPS and is typically around 30%. In such cases efficiency of the UPS at 30% load is what should be considered.

Efficiency of a UPS is dependent on the technology used in the power transistors in the switching circuits. High efficiency transistors such as IG-BTs improve the efficiency of switching. Also, advanced Digital Signal Processing based switching controls improve the efficiency of the UPS. The UPS design topology, i.e. the way the internal components are connected within the UPS, also impacts the efficiency. High efficiency topologies such as "Delta Conversion Online", improves the efficiency by reducing the no load and square law losses.

There have been reports of extreme UPS design where the UPS is made part of the server itself. A battery is integrated with the server that allows it to function as a UPS. In such cases it has been possible to achieve over 99% efficiency, as compared to a maximum of 95% efficiency in conventional UPS systems.

6.3.4 Servers

It is projected that by 2020, servers will be responsible for more than half of the total data center CO_2 footprint – which is close to 10% of all ICT footprints [4]. The bulk of this is from low cost commodity servers based on x86 processors from Intel or AMD. Servers, especially commodity servers, therefore demand close attention of the ICT community from green ICT perspective.

Figure 6.5 shows the distribution of peak power consumption for components of a typical commodity server. This distribution gives an idea where the power goes in a server. Maximum power reduction can be achieved by increasing the energy efficiency of large power consuming components like CPU, memory and power supplies.

We now take a deeper look into some of the best practices for reducing server power consumption.

Powering of Unused Servers

Powering off idle servers in the data centers is the simplest way of reducing power consumption. In many data centers there are unused servers – for example servers that are still being operated for projects that have been completed. Such servers consume power and space but produce nothing useful as

Server Power Consumption

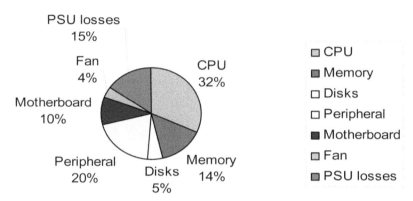

Figure 6.5 Distribution of server power consumption

output. By some estimates there are close to 5 million unused servers in the world that consume around $4 billion worth of energy [14]. It is important to note that server utilization is not the same as CPU utilization. It is easy to measure and track CPU utilization but it is more difficult for data center managers to get a visibility of the useful work performed by a server. The CPU of an unused server will show activity even when it is doing housekeeping tasks but such CPU activity does not add any business value. Effective inventory tracking and configuration management practices in the data center help in tracking server usage and in identifying unused servers.

Server Consolidation

Once unused servers have been eliminated, the biggest gains in energy efficiency are achieved via server consolidation [15]. A large majority of servers in most data centers consist of low cost commodity servers that run single applications. Typical CPU utilization levels of x86 servers are between 5 to 15%. Since servers are not energy proportional, the underutilization of servers results in massive energy inefficiency. Server consolidation helps improve the utilization levels of all components of the server (CPU, network and I/O). The following are some the ways for consolidating servers:

- Creating multiple hardware partitions of single server hardware is possible on mid-range and high end servers. Each partition is electrically isolated and contains its own CPU, memory, network, and cache and can run an operating system image. Examples include Sun's Dynamic System Domains available on UltraSPARC servers.

- Replacing many low power servers with a single more powerful server with multiple cores/CPUs/hardware partitions
- Running more applications on a single server – for example running multiple web applications on a single web server
- Running multiple customers/user groups on a single instance of application

Consolidation Using Virtualization

Virtualization software such a hypervisors enable multiple operating systems instances to run on a common shared server hardware in such a way that each operating system has the illusion that it has its own dedicated hardware. Virtualization of multi core x86 bases servers using hypervisors like VMware and Xen has been widely adopted by data centers and is now one of the most effective ways for consolidation and increasing CPU utilizations. Users have achieved CPU utilization levels in the 70–80% range using hypervisors. This is possible by running as many as 30 to 40 Operating System instances on the same hardware. Adoption of virtualization has resulted in dramatic reduction in the number of servers in the data center and corresponding savings in space, powering and cooling expenses.

Energy Efficient Servers

The server hardware itself must be energy efficient. Energy efficient servers are designed for reducing power consumption using techniques such as

- Energy efficient silicon technology.
- Power management capabilities in the CPU that enable the CPU to enter various power down states depending on workload.
- Minimizing waste in power supplies and power conversion electronics.

Energy efficient silicon in server CPUs contributes significantly towards reduction of server power consumption. One of the biggest components of power consumption in the CPU is the leakage power that results due to the leakage current flowing through silicon transistors when they are in the off state. New materials known as "high-k" materials are being used in place of Silicon Dioxide (SiO_2) as the transistor gate dielectric material. These materials are able to reduce the gate leakage by a factor of about 100. As a result the heat loss of such chips is much lower and the devices run at lower temperatures.

Till 2002 performance of CPUs was largely governed by increasing the clock frequency. Since then, we have seen the introduction of multi-core CPUs that have multiple processing engines each of which run at lower voltage levels and lower clock frequency. This results in larger aggregate performance but at substantially lower power consumption. Studies have shown that quad core Intel Xeon processor based servers in 2008 gave seven times the performance for the same power consumption as compared to first generation single core Xeon processor based servers of 2004.

The idle power consumption for new generation CPUs is lesser than older generation processors. Internal measurements by Intel have reported 50% less idle power consumption for the latest generation of Xeon 5500 quad core processor based servers as compared to previous generation of processors [16].

Power Management and Power Reduction

Power management and power reduction involves careful monitoring of the system level activity and then controlling the power consumption as per workload. This is done both at the server level and at the CPU level.

At the CPU level, it is possible to put the server in various "power down" states. A modern x86 based processor such Xeon 5500 has around 5 power/sleep states. When a "deep sleep" state is invoked, the CPU power consumption is reduced significantly. However, each sleep state is also associated with certain latency and each state transition itself consumes energy. Operating System kernel and device drivers may take advantage of these sleep states to optimally reduce the aggregate power consumption. Similarly, applications must be made "power friendly" so that they enable the operating system to operate the CPU at the most optimum state. LessWatts – an open source project has reported a 4.5% reduction of idle power consumption on a Xeon 5300 server using a power friendly Linux kernel [17].

Server Power Management software enables measuring of server power usage and provides visibility to data centers managers. Power management software allows dynamic capping of server power consumption to preset levels. It is possible to control power consumption by dynamically adjusting processor clock speeds based on actual server load and power consumption [18].

Cloud Computing

With the emergence of cloud computing infrastructure services from companies like Amazon and Microsoft, there is an option of moving corporate data center functions to the cloud. Cloud infrastructure providers run huge data centers at extremely high PUE levels, often less than 1.2. It is not possible for even large sized corporate data centers to reach the scale of cloud infrastructure. Cloud providers will therefore always be able to run infrastructure with higher efficiency than corporate data center by using techniques mentioned in this chapter and by leveraging scale. Therefore, a green option for the corporate customer would be to host applications on the cloud instead of running one's own data center.

6.3.5 Storage

Storage systems are responsible for 6–7% of the data center carbon footprint. In a given data center however the power consumption of storage devices would vary depending on the applications, design and I/O profile. Storage is the one of the fastest growing segments of the IT market with ever increasing demand for storage capacity because of rapid growth from applications such as multimedia, high definition television, user generated content, VOIP, security & surveillance and email.

We now discuss some of the issues that affect storage power consumption.

Reducing Disk Power Consumption

Power consumption of disk depends on the disk diameter, number of disk platters and rotational speed and is given by the following relationship:

$$\text{Power consumed in a disk} \sim \text{Diameter}^{4.6} \times \text{RPM}^{2.8} \times \text{Number of Platters}$$

Therefore for reducing power consumption of disk drives one should

- Choose the optimum rotational speed. Lower rotational speed comes at the cost of increased I/O latency for random access. For example for the Hitachi Ultrastar 3.5 inch disk – the 15k RPM 450 GB disk has an idle power consumption of over 13W, but the 7.2k RPM 500 GB disk consumes 7.3W when idle. However, for sequential read access, rotational speed is less of an issue since we are more concerned about disk bandwidth.
- Choose the right disk size. A 2.5 inch disk will consume less power than a 3.5 inch disk.
- Choose the right capacity. The disk capacity depends on the number of platters and disk diameter. For the same disk diameter, a higher capacity disk – say a 1 TB disk would consume more power than a 500 TB disk. For example, for the Hitachi Ultrastar 3.5 inch 7.2k RPM disk, the 1 TB

disk has 5 platters and consumes 9 W when idle and the 500 GB disk consumes 7.3W when idle.

Some metric used for comparing energy efficiency of storage devices is Watts/GB and Watts/IO operations/second. For IO bound applications the former should be used. However, large capacity disks typically provide better performance on a Watt per GB metric.

Powering Down Disks

Disks exhibit highly non-proportional behavior since even when no data is being accessed they continue to rotate. Recent developments in disk drives include multiple "power down" modes similar to CPUs. Low RPM idle modes have been designed that reduce the power idle power consumption by almost 50%. Most modern disk drives have automatic power management features that optimize the idle power consumption. Moreover disks have standby and sleep modes that reduce the idle power to even lower levels. Bulk storage systems such as MAID – Massive Array of Idle Disks employ disk spin down to save on power. In these systems disks are kept in spun down state. Here power saving is achieved at the cost of I/O latency, the time that is required to spin the disks up.

Solid State Storage

Solid State storage devices are based on NAND flash technology and do not have any mechanical component. Solid state storage devices are expensive from a $/GB perspective but they are most efficient from an I/O performance perspective. A solid state storage device such as Fusion IO provides more than 100,000 I/O operations per second at less than 10 Watts of power consumption. Such devices can be used in storage designs where large number of disk drives is used only to increase the I/O performance. To produce an equivalent IO performance using the best quality disk drive would require 2000 times more power.

Solid State storage is also being used in place of boot disks in servers. Such servers consume much less power and give better performance than disk based systems. However, solid state disks are about 40 times or more expensive than ordinary disks on a per GB basis.

Power Efficient Storage System Design

Just as in the case of servers, effective storage management practices such as storage virtualization, thin provisioning, data compression and de-duplication results in more effective use of storage infrastructure [19]. This allows users to avoid or delay purchase of additional storage systems and also save power.

The choice of RAID in a storage system also has a bearing on the power consumption. In RAID arrays, data is either mirrored on additional disks or parity information is written on additional disks. This improves the reliability and availability of the disk storage system. However, this comes at the cost of additional power consumption. If energy consumption is an issue, RAID may be avoided if possible or the lowest level of RAID may be used to reduce power consumption.

6.3.6 Network

The share of network equipment in the total data center power consumption is much less than servers. Still, there is a need to reduce the power consumption of network equipments like switches and routers. Energy efficiency of network equipment is measured in terms of Watts/Gbps of network traffic.

Basic techniques for improving the energy efficiency of switching and routing devices are similar to those for servers. Use of improved silicon, improved power supply efficiency, improved power management by leveraging power down states and improving thermal and cooling designs are resulting in more energy efficient switches and routers.

Emergence of high bandwidth network technologies such as 10 Gb Ethernet and Infiniband offer further scope for reducing energy consumption in the data center. It is possible to consolidate multiple gigabit network interfaces and Fiber Channel storage network interface on a single 10 Gb Ethernet host bus adapter using virtualization techniques. It is thus possible to consolidate both data and storage network on a single unified data center Ethernet switching fabric based on 10 Gb Ethernet. This benefits the data center by simplifying the network, reducing the number of devices and networks and reducing the count of Host Bus Adapters needed in the servers. All this leads to reduction in power consumption in the data center.

Energy Efficient Ethernet

Considering the fact that Ethernet dominates data center networks, office networks and now even homes, a need was felt for improving the energy efficiency of Ethernet networks. IEEE has formed a task group – IEEE 802.3az Energy Efficient Ethernet Task Force to create a standard and bring it under the umbrella of 802.3. Energy Efficient Ethernet [20] aims to reduce energy consumption during periods of low link activity. The central idea is that the

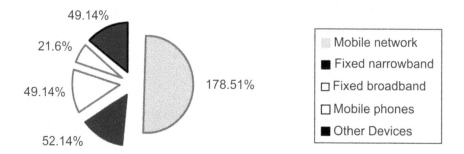

Figure 6.6 CO_2 footprint of telecom devices and infrastructure (adapted from [4])

link should switch to a lower speed and hence lower a power PHY when link utilization is low. It has been seen that in a network switch, as the number of active links increase, the power consumption shows high levels of non-proportional behavior. Considering the fact that that at lower link speeds, significant power is saved, significant energy efficiency is achieved by switching to lower link speeds.

6.4 Green Telecom Infrastructure and Devices

As much as 25% of the total ICT footprint will be from telecom devices and infrastructure. This amounts to almost 349 million tons of CO_2 equivalent [4]. Figure 6.6 shows the breakdown into various segments. In this section we look into details of energy consumption in these segments and best practices for reducing carbon footprint.

6.4.1 Mobile Networks

Mobile network equipment is operated nonstop round the clock and 365 days a year. As the number of mobile subscribers increase, a higher number of cell sites is added to the network and the energy bill for maintaining the network continues to soar. The share of energy consumption and emission by large

Table 6.1 Efficiency level of RBS components

Component	Efficiency
Feeder	~50%
Radio Equipment	~6%
Power Supply	~85%
Air Conditioning	~35%

operators in a country is indeed large. In 2008, the total CO_2 emission from Verizon amounted to a staggering 6.2 million metric tons of CO_2 [21].

Almost 80% of a mobile operator's energy consumption is due to radio base station equipment. The remaining is consumed in core networks. Let us now look at approaches for reducing energy consumption in mobile network.

Network Optimization

An optimal network design is the first and most important step in achieving energy efficiency [22]. A poorly designed network will result in a sub optimal selection of number and type of radio sites. A good design can lead to 30 to 50% reduction in number of radio sites and bring about substantial reduction in energy consumption, even with the same level of efficiency as far as the radio components are concerned. The designer must start with a careful assessment of capacity, coverage and QoS requirements. The design must also take into consideration future growth requirements and presence of competing networks in the location.

Once the requirements are captured, the design can be optimized with respect to number of cell sites while meeting the requirements. A good design results in reduced number of cell sites and corresponding reduction in TCO (Total Cost of Ownership) over the lifetime of the radio sites. This would include power costs, rentals, maintenance and transportation.

Site Optimization

The radio base stations site and equipment optimization would include maximizing energy efficiency of the radio equipment, signal processing and associated circuitry, power supply efficiency and air conditioning [23]. Figure 6.7 shows the breakdown of power consumption in a traditional base station.

If we take the RF load as the final useful output from the base station, we find radio base stations are work at efficiency levels ranging from 1–4%. At the component level, a typical radio base station operates at the following approximate levels of power efficiency.

Base Station Power Consumption by Function (% of Total)

Figure 6.7 Breakup of power consumption in a typical radio base station site (adapted from [23])

Any efficiency gained at the end of the RF power chain has a cascading effect on the total power consumption. One Watt energy saved in the RF level translates into a total of 28 Watts saved for the total base station. The following are some of the methods being adopted for saving power:

- Use of Remote Radio Units (RRU) – Significant inefficiency happens because of radio feeder losses. It is more efficient to move the radio power amplifiers and converters to the base of the tower in a separate remote radio unit and thereby eliminate the feeder path and associated losses. Optical Fiber cables can be used to carry signals from the signal processing units that continue to remain in the base station rack. It is estimated that this alone would bring in 40 to 50% efficiency gains compared to current practice.

- Radio Standby Modes – Power saving is achieved by judicious use of power down states. Typical base stations have three sectors cells with each sector having about four transceivers. During periods of low activity, such as during the night, some transceivers in each sector can be put in standby mode and thus save substantial power. Power savings of 10–15% are possible without any degradation of service.

- Power Supply Efficiency – Similar to data centers, electrical system efficiency in a radio base station can be improved by using higher efficiency rectifiers in the AC to DC power conversion process. Efficiency is improved when rectifiers are operated at higher load levels. A base station may have as many as five rectifiers or more. Power management and automatic control systems can be used to intelligently vary the duty cycle of rectifiers so that some are in power down state and others are operated at higher load levels as per actual load conditions.

- Cooling system efficiency – Similar to data centers, "free cooling" using outside air may be used in base stations.
- Efficient operation of remote sites – Radio sites in remote locations that are not on the electricity grid are usually run on redundant pairs of generator sets. A more optimal solution is to replace one generator with a battery pack and use the generator for charging the battery. During night time when the base station requires less power, the battery pack is used. During day, the battery is charged using the generator which is run at full load and hence at higher efficiency. The battery may also be charged using other sources such as solar and wind power.

Spectrum Issues

Spectrum efficiency and energy efficiency are partly contradicting requirements. In general, lower frequencies provide better coverage using the same amount of power transmitted since lower frequencies have better penetration capabilities through walls and vegetation. Use of lower frequency may allow wider coverage with fewer radio sites and hence lower energy costs.

6.4.2 Use of Alternative Sources of Energy

Use of alternative energy is important because of multiple reasons. On one hand use of alternative sources is eco-friendly and contributes towards the top line of the network operator [22]. On the other hand, it enables powering of radio sites that are off the electricity grid. Table 6.2 lists some alternate sources of energy and the corresponding opportunities and challenges.

6.4.3 Wireline Networks

Energy consumption in wireline networks is primarily driven by the proliferation of broadband subscribers. It is estimated that there are close to 450 million broadband subscribers in the world using broadband access technologies such as DSL, Cable and FTTP [24]. This number in turn is driven by number of PCs in the world, currently above 1 billion. Broadband equipment by definition is any access device that is capable of providing at least 2 mbps full rate data transfer in at least one direction. It is projected that there will be close to 900 million broadband routers by 2020.

Examples of equipment covered under broadband include

Table 6.2 Alternate sources of energy for Radio Base Stations

Source	Advantages / Opportunity	Challenges
Solar Power	• Low opex • Usable for small to medium capacity base stations • Takes advantage of energy efficient base stations • Low maintenance costs	• High capex • Requires more space • Requires more energy storage capability to cater for seasonal and climatic variations in solar power
Wind Power	• Low Opex • Can support larger radio sites	• Erratic supply • Requires other sources like generators, etc., for backup • Extra space required
Fuel Cells	• As a replacement for battery and generator sets	• Technology is not very mature • Re-fuelling is a challenge
Bio Fuels (e.g. bio diesel)	• Cheaper than normal diesel • Lower carbon emission, lead and sulfur free and non toxic	• Need to follow environmental guidelines, asses overall savings and environmental impact

- Home Gateways – e.g. customer premises DSL routers, cable routers, optical network termination units, Ethernet routers and Wi-Max routers.
- Set Top Boxes for IPTV.
- Simple CPE devices – e.g. USB connected modems.
- Home infrastructure devices – Wi-Fi access points, power line adapters, hubs and switches.
- Network Service provider equipment – e.g. DSL equipment, optical network termination equipment, wireless broadband equipment such as Wi-Fi hotspots and Wi-Max radio, Cable service provider and Power line service provider equipment.

According to an estimate Europe may consume up to 50 TWh of electricity for powering up broadband equipment. To promote responsible behavior from all stakeholders, the European Commission has come out with a Code of Conduct [25] for equipment and component manufacturers, service providers and network operators for adoption of energy efficient broadband devices.

The code of conduct has identified three different power levels for customer premises broadband equipment – namely On, Off and Low Power. Similarly for service provider end network equipment the three states defined are Full Power, Low Power and Standby. The code of conduct also lays down the targets for power consumption in each level and a schedule for achieving the same.

The description of the power states are as given in Table 6.3.

Table 6.3 Power states for broadband equipment

Class	Power State	Description
Customer Premises Equipment	On	All components are in on state and device provide at least 25 % of full bandwidth in all directions
	Off	Device is switched off or not connected to mains and does not provide any functionality
	Low Power	Device is idle, components are in low power state, device is not processing any traffic but is ready to detect activity
Network Equipment	Full Power	The full power state where full data transmission is allowed
	Low Power	Limited power reduction is achieved and limited data transmission is allowed
	Standby	Lowest power state where no data transmission takes place, device is able to respond to activation request and changes state

Table 6.4 European Commission Code of Conduct. Example targets and schedule (Source [25])

Equipment	Low Power state (W)		On state (W)	
	2009/2010	2011	2009/2010	2011
ADSL home gateway with 4 Fast Ethernet ports, a single radio 802.11b/g Wi-Fi interface and 2 USB ports	6.6	4.4	9.8	8.1
VDSL2 home gateway with 4 Fast Ethernet ports, a single radio 802.11n Wi-Fi interface, 2 USB ports and 1 FXS port	8.3	6.1	14.3	12.3
Fast Ethernet router with 1 WAN and 4 LAN Ethernet ports	3.7	3.1	6.4	5.1
Wi-Fi Access Points with single band IEEE w802.11b/g or 11a	3.0	2.3	4.0	3.6
ADSL-2 Network Equipment – per port	1.1	0.8	1.3	1.2

Some example targets for home gateway devices and schedule set by the Code of conduct are given in Table 6.4.

Currently, efforts are made to reduce the energy requirement of wireline equipment to meet or exceed the above targets. The most promising technology looks to be GPON – Gigabit Passive Optical Networks which is an optical communication technology. Fiber technology increases the bandwidth and range and reduces the energy consumption at the same time. A GPON system uses 80% less energy as compared to ADSL.

6.4.4 Mobile Devices

Although the share of the carbon footprint of mobile devices themselves is much less than that of mobile and wireline networks, considering the sheer number of mobile devices, the absolute value is not trivial. In 2008, an estimated 1.2 billion mobile phones were sold worldwide. By 2020, mobile devices themselves will emit about 21 million tons of CO_2 annually. The biggest contributor to the carbon footprint is due to the power that is consumed when the mobile charger is kept switched on even after the mobile battery has been completely charged.

The US Environmental Protection Agency has set targets for mobile battery chargers to meet the requirements of the Energy Star label. The World GSM Association has initiated the development of a standard for mobile chargers called the Universal Charging Solution [26]. This standard aims to reduce standby energy consumption, eliminate the proliferation of duplicate chargers and enhance the mobile user's experience. GSMA expects that UCS will eliminate about 51000 tons of duplicate chargers. This charger will use a detachable micro USB cable and will have a no load power consumption less than or equal to 0.15 W.

6.4.5 Impact of Next Generation Networks

As the world moves towards the Next Generation Networks (NGN), the network architecture that will unify the fixed, mobile and broadcast networks, significant energy efficiencies will be achieved [27]. NGNs are characterized by a common shared network that uses end to end IP based packet switching. NGNs use common platform and service oriented infrastructure for offering user centric services on the converged network. NGNs use a common soft switch platform for both connection oriented and connectionless communication.

NGNs are energy efficient because of the following reasons:

- Significant transmission efficiency is achieved because of use of IP based communication. Use of digital compression increases the capacity of network. This leads to better utilization of the infrastructure and hence energy.

- NGNs enable the use of power down modes for various customer premises and service provider equipment.
- Optical transmission is a key ingredient in many NGN implementations. Use of optical networking, especially Passive Optical Networks results in significant energy savings when compared to copper based networks for the same amount of data transfer.
- Traditional network architecture requires separate switching and routing infrastructure for each service. With NGNs, a common pool of switching centers can handle all services. The number of switching centers can be dramatically reduced.
- Migration of separate networks on a single shared IP network enables consolidation of applications and support infrastructure on common platforms. It is now possible to have a common Operational Support System and Business Support System in place. This reduces not only the number of elements in the infrastructure, but also simplifies the day to day operation and management. Data center and other IT assets are also shared and better utilization leads to energy efficient operations.

6.5 Impact of PCS and Peripherals

By far the largest contribution to CO_2 emissions in the ICT sector is from PCs and peripherals [4]. This is due to the fact that there are already more than a billion PCs worldwide and the number is expected to touch 4 billion by 2020. PCs alone will be the single largest contributor to ICT emissions, responsible for almost 42% of all ICT emissions.

By 2020, two major technological changes are expected to take place. First, desktop PCs will be largely replaced by the more energy efficient laptops. Secondly, almost 100% of CRT displays will be replaced by energy efficient LCD displays. Both these changes will bring in significant efficiency improvement. However, the sheer increase in number of machines will mean that the total CO_2 footprint in 2020 will be three times that of 2002 levels.

In this section we look at the components of PCs that deserve special attention with respect to energy efficiency – namely displays, power supplies and disks. We look at how intelligent power management techniques can be used to reduce PC power consumption in office environments.

We start with a brief note on the Energy Star* Program and what it means as far as energy efficient PCs and laptops are concerned. Energy Star* is a voluntary program jointly administered by the US Environmental Protection

Agency and the US Department of Energy. Energy Star* aims at reduction of CO_2 emissions from a wide gamut of products including household appliances, heating and cooling systems, home electronics, lighting, computers, peripherals and many more classes of products. Products that meet Energy Star* requirements get an Energy Star* label that helps consumers with selecting energy efficient products. Energy Star* Version 5 has been released in July 2009 and covers product like computer displays, power supplies, desktops, laptops, thin clients and work stations.

Computer Displays

Energy Star* display specifications covers products such as computer displays, digital picture frames and digital signage. Energy Star* specifies strict power target levels for various power down states – namely On, Sleep and Off modes. In sleep mode, displays must consume less than or equal to 2 W and in off mode less than or equal to 1 W. For computer display size less than 30 inch diagonal, the on mode power consumption is based on a formula that factors in the screen size and display resolution. Energy Star* displays are about 25 to 60% more efficient than non-qualified displays.

Power Supplies

The Energy Star* Version 5 [28] specifies that power supply for PCs must be at least 85% efficient at 50% of the rated output and 82% efficient at 100% and 20% of the rated output. Moreover, the power factor must be at least 0.9 at 100% of the rated output capacity. Power supply efficiency typically falls at a rapid rate when the output is below 20% of the rated capacity and can become lower than 60%. This means that when the PC is idle, there are significant power losses due to power supply inefficiency.

Desktops vs. Laptops

Laptops are much more energy efficient as compared to desktops. Significant energy saving may be achieved when an organization replaces desktops with laptops. The Energy Star* Version 5 specifies Typical Energy Consumption (TEC) targets for desktops and laptops. TEC is a measure of the annual power consumption of a desktop or laptop based on typical usage pattern that includes periods of Off, Sleep and Idle. The targets are specified for various categories of laptops and desktops based on factors such as number of cores of processors, memory, graphics processor etc. Comparing the targets for laptops and desktops, we find that the TEC target of the highest category of laptop is less than 88.5 kWh whereas even for the lowest category of desktop it is 148 kWh.

Solid State Disks

Solid state drives consume less energy than hard disk drives. The average active and idle power is lower than hard disk drive by an order of magnitude. In case of laptops, this translates into longer battery life and lower temperatures.

Energy Efficient Silicon

Just as in the case of servers, desktops and laptops benefit from use of modern multi-core processors that run at lower voltage levels and lower frequencies. A desktop or a laptop based on energy efficient silicon will result in significant power savings.

Power Management

Power management techniques such as reducing the use of the maximum power scheme, adaptive CPU throttling to reduce clock speeds during idle use of standby modes help reduce power consumption. Also, technologies such as Wake on LAN can be used by companies to remotely power down systems at night and then again power them up in the morning.

6.6 Global Efforts in Green ICT

At the global level, several initiatives and programs are on the way to the "greening" of ICT and leveraging the power of ICT for creating the low carbon economy and society. In this section we take a look at only a few of the important organizations and programs active in this space.

Global e-Sustainability Initiative (GeSI, http://www.gesi.org)

GeSI is a non-profit organization, headquartered in Brussels, Belgium, that brings together ICT companies, industry associations and NGOs to further the cause of sustainable economy using innovative use of ICT. GeSI was formed 2001 and has over 25 global companies and organizations as its associate members. Some of the world's largest technology companies and service providers are its members. In 2008 GeSI published a seminal work – the *Smart 2020 Report* [4] on use of ICT for creating a low carbon economy. GeSI's activities include development of standards, methodologies, best practices and promotion of good conduct in the areas of Climate Change, Energy

Efficiency, E-waste and Supply Chains. It also works with policy makers to promote the use of ICT for sustainable development.

The Green Grid (TGG, http://www.thegreengrid.org/)
The Green Grid is a global consortium of over 180 organizations world-wide. It was formed with some of the world's technology companies as its board members. The Green Grid focuses on development of promotion of energy efficiency in data centers and enterprise computing environments. Green Grid's charter includes defining meaningful, user-centric models and metrics for data center efficiency. It also promotes the adoption of energy efficient standards, processes, measurement methods and technologies. The Green Grid is best known for publishing the PUE/DCiE and DCP metrics for data centers.

Climate Savers Computing Initiative (CSCI, http://www.climatesaverscomputing.org/)
The Climate Savers is a non-profit organization started in 2007 by Google and Intel. It has as its board some of the world's leading PC technology companies like Dell and HP. It has over 500 members worldwide. The focus of this group is on energy efficient PC and servers. Climate Savers promotes the development, deployment and adoption of energy efficient technologies for PC power delivery and use of power down states for reducing the idle state energy consumption. The stated goal of CSCI is to move industry toward a 50% reduction in power consumption by computers by 2010.

US Energy Star* Program (http://www.energystar.gov/)
As mentioned in Section 6.5, Energy Star* is an initiative of the US Envir-onmental Protection Agency and Department of Energy. In 1992 the EPA established Energy Star* as a voluntary program, called Energy Star to cover first computers and later other categories of office equipment and other products. The Energy Star program for computers aims to generate awareness of energy saving capabilities and accelerate the market penetration of more energy-efficient technologies. In July 2009, Energy Star* releases the Version 5 of its specifications for PCs, laptops, power supplies, displays, workstations and thin clients.

Smart Grid Interoperability Panel (SGIP, http://collaborate.nist.gov/twiki-sggrid/bin/view/SmartGrid/SGIP)

The Smart Grid Interoperability Panel (SGIP) is a membership-based organization created by the US National Institute of Standards and Technology (NIST). NIST is responsible for coordinating the development of and publishing a framework, including protocols and model standards, to achieve interoperability of Smart Grid devices and systems. SGIP has is an Open, transparent body that has representatives from all Smart Grid stakeholders – such as government agencies and private members. SGIP does not directly develop or write standards. It helps in ther coordination, acceleration and harmonization of standards development. SGIP reviews use cases, helps identify requirements and coordinates conformance and interoperability testing, and creates action plans for achieving interoperability.

Initiatives of the European Commission (EC, http://ec.europa.eu/index_en.htm)

EC has acknowledged the importance of the use of innovative ICT-based technologies for achieving a low carbon emission targets in a cost effective manner. In October 2009 it issued a set of recommendations for its member states for mobilizing ICT for transition to a low carbon economy. The recommendations include leveraging the following ICT capabilities:

- Use of modeling, simulation, design, monitoring and visualization tools for smart buildings.
- Use of ICT for functioning and optimization of transportation and logistics operations.
- Use of ICT based solutions for measuring, monitoring and managing carbon emission in all energy using activities across sectors.
- Dematerialization and electronic service delivery.
- Use of ICT in creating smart grids, deployment of smart meter infrastructure for real time energy use monitoring, enabling better management and control of energy and reducing the final energy consumption of end users.

The Commission is promoting R&D projects for developing ICT tools to improve energy efficiency. Under the Seventh Research Framework Programme (FP7), it supports projects that aim to develop energy-positive buildings and neighborhoods (which generate the energy they need and sell any surplus generated), and developing the smart energy grid. One such project is the

SmartGrids European Technology Platform for Electricity Networks of the Future (http://www.smartgrids.eu/?q=node/27).

6.7 Standards and Green ICT

This section lists some of the key areas where ICT standardization will impact climate change mitigation efforts and how GISFI can contribute.

Technology Standards

Standards will play a major role in CO_2 emission mitigation and climate change efforts [29]. ITU has already set up the ITU-T Study Group 5 on Environment and Climate Change. The study group has started work in areas such as

- Coordination and planning of ICT standardization related to climate change.
- Methodology for environmental impact assessment for ICT.
- Energy efficient power feeder/supplies for ICT devices.
- Methodology for data collection related to ICT energy efficiency.
- Environmental protection and recycling of ICT equipment and facilities.

Indian participation in this study group must be encouraged and harmonization of related standards for the Indian scenario is required.

Some of the technology areas where ICT standardization will influence climate change mitigation efforts are given below. In each of these areas Indian participation in relevant world standard bodies must be encouraged.

- Next Generation Networks – Impact of NGNs have been discussed in Section 6.4.5.
- Internet of Things – Under the IoT umbrella we consider the following two technologies: Ubiquitous Sensor Networks (USN) and Radio Frequency Identification (RFID). Ubiquitous Sensor Networks built using wired and wireless sensor networks (WSN) can be used in climate and environment monitoring, emission measurement and monitoring, marine environment and glacier monitoring, forest and vegetation monitoring, home automation and industrial control. RFID technology will be used in smart logistics and energy efficient supply chains.
- Smart Grid Interoperability – Standardization is essential in achieving interoperability of Smart Grid technologies such as Smart Meters, Advanced Metering Infrastructure and Home Area Networks (HAN) based smart energy management systems. In this regard GISFI needs to work

in coordination with the government of India, Department of Power and contribute in its effort in modernizing India's power sector. GISFI may also participate the Smart Grids Interoperability Panel.

- Intelligent Transportation Systems – Standardization and interoperability is required with respect to ubiquitous connectivity, vehicle data collection, navigation, telematics, etc.

Radio-Communication Standards

Radio-communication and spectrum issues also have an impact on climate change efforts. How spectrum issues impact energy consumption of radio base stations has been discussed in Section 6.4.1. Apart from this, spectrum is needed for use in meteorological monitoring and measurement systems and USNs as discussed above. Finally, there is a need to look at standardization of new radio technologies like digital modulation, compression and error correcting codes that will enable radio equipment to lower their energy requirements.

Other Areas Deserving Focus

GISFI can contribute by having specific focus groups and programs in areas such as

- Awareness Building and Knowledge Dissemination: Indian ICT industry needs to be educated about the impact of ICT on CO_2 emissions – both the negative impact and how ICT can be used to reduce emissions. India already has close to 500 million mobile phones and hundreds of millions new phones will be added in the coming years. There is a huge carbon footprint in this respect and it is high time industry starts measuring emission and initiates necessary action.
- Energy labeling programs: There is a need to enhance existing Energy Labeling programs in India – such as the ones run by the Bureau of Energy Efficiency, Government of India [30] – to include ICT products.
- Metrics programs: There is a need to come out with metrics and performance indicators for energy efficiency of all relevant ICT activity in India. This includes harmonization of global best practices and guidelines for India. Also, there is scope to develop new guidelines for areas in which there are no global standards or best practices so far, for example energy efficiency metrics for Radio Base Stations.

- Code of Conduct: There is a need for establishing Code of Conduct for Indian ICT operations similar to the European Code of Conduct for achieving time bound targets of energy efficiency.

6.8 Conclusions

Climate change is one of the biggest challenges that humanity is facing today. The global community has no option but to mitigate the ill effects of atmospheric greenhouse gases by cutting down on carbon emission intensity. ICT technology offers cost effective ways of reducing the carbon footprint of various sectors of the economy and helps in creating a sustainable low carbon future. ICT enabled Smart Energy Grids, Smart Buildings, Smart Factories, Smart Logistics and Smart Transportation can help bring down emission by 15% from BAU levels.

The ICT sector itself emits carbon. PCs and peripherals, mobile networks and data centers are the biggest contributors of carbon emission in the ICT sector. Measurement and monitoring of energy used, consolidation and principle of energy proportionality are generic approaches to minimizing energy consumption in the ICT sector.

Energy efficiency of data centers can be substantially improved by improving the efficiency of cooling systems. Use of "free cooling" techniques such as outside air help reduce the cost of cooling. Electrical distribution systems efficiency is improved by using high efficiency components and running the systems at optimum loads. Consolidation and virtualization techniques help improved CPU utilization and lower energy costs.

Mobile networks consume a lot of energy. Optimal network design and efficient radio base stations can reduce power consumption substantially. Intelligent power management can help reduce power consumption further. Broadband equipment must support intelligent use of idle and low power states. Home gateways and modems must be made highly energy efficient since their numbers are going to explode. Optical networks will provide the most energy efficient networks. In case of mobile devices, it is important to reduce the power consumption of idle battery chargers.

PCs and peripherals will have the largest CO_2 footprint in the ICT sector since the number of PCs is expected to touch 4 billion by 2020. Efficient use of power management and reduction of idle power will help in reducing power. Laptops are more energy efficient than desktops and may eventually replace most PCs.

There are several global initiatives in progress in the area of green ICT. Standards have a big role to play in the adoption of green practices in ICT and deployment of ICT for reduction in carbon emission levels. GISFI as a standardization forum has an important role to play in the Indian context.

References

[1] D. Papadimitriou, Future Internet: The cross-ETP vision document. European Technology Platform, Alcatel Lucent (ed.), 8 January 2009.

[2] Department of Telecom, Government of India, Memo for Telecom Commission, http://www.dot.gov.in/network/2009/stt1.pdf, November 2009.

[3] Europe's Digital Competitiveness Report – i2020 – ICT Country Profiles, Commission of the European Communities, EU Report, August 2009.

[4] The Climate Group: Global e-Sustainability Initiative report: Smart 2020 Enabling Low Carbon Economy in the Information Age, http://www.smart2020.org/publications/, 2008.

[5] L. Barroso and U. Holzle, The case for energy proportional computing, Google, IEEE Computer, December 2007.

[6] PUE and DCiE, The Green Grid Data Center Power Efficiency Metrics, http://www.thegreengrid.org/en/Global/Content/white-papers/The-Green-Grid-Data-Center-Power-Efficiency-Metrics-PUE-and-DCiE, October 2007.

[7] Report to Congress on Server and Data Center Energy Efficiency Public Law 109-431, U.S. Environmental Protection Agency Energy Star Program, http://www.energystar.gov/ia/partners/prod_development/downloads/EPA_Datacenter_Report_Congress_Final1.pdf, August 2007.

[8] M.K. Patterson, The effect of data center temperature on energy efficiency, Intel Corporation, http://ieeexplore.ieee.org/iel5/4538254/4544243/04544393.pdf?arnumber=4544393.

[9] Intel Information Technology, Reducing data center cost with an air economizer, http://www.intel.com/it/pdf/reducing_data_center_cost_with_an_air_economizer.pdf, August 2008.

[10] The Hot Aisle, IBM claim that water cooled servers are the future of IT at scale, http://www.thehotaisle.com/2009/06/03/ibm-claim-that-water-cooled-servers-are-the-future-of-it-at-scale/.

[11] 2008 ASHRAE Environmental Guidelines for Datacom Equipment, http://tc99.ashraetcs.org/documents/ASHRAE_Extended_Environmental_Envelope_Final_Aug_1_2008.pdf.

[12] N. Rasmussen, AC vs DC power distribution for data centers, APC Whitepaper, http://www.apcmedia.com/salestools/SADE-5TNRLG_R5_EN.pdf.

[13] R.L. Sawyer, Making large UPS systems more efficient, APC Whitepaper, http://www.apcmedia.com/salestools/VAVR-6LJV7V_R1_EN.pdf.

[14] A. Bednarz, Unused servers squander $25 billion a year, CIO Magazine, `http://www.cio.com/article/505351/Unused_Servers_Squander_25_Billion_a_Year`, October 2009.

[15] B. Baikie and S. Gaede, Easing the space, power, and cooling crunch with green Internet datacenter re-design, Sun Microsystems White Paper, Second Edition, January 2008.

[16] Intel Xeon Processor 5500 Series Product Brief, `http://www.intel.com/Assets/en_US/PDF/prodbrief/xeon-5500.pdf`.

[17] LessWatts project – Saving power with Linux, `http://www.lesswatts.org/`.

[18] Charles Lefurgy et al., Server-level power control, IBM research `http://www.research.ibm.com/people/l/lefurgy/Publications/icac2007publish.pdf`.

[19] SNIA Tutorial, Green storage: The big picture, SW Worth, `http://www.snia.org/education/tutorials/2009/fall/green/SWorth-GreenStorage_SNW-Oct-09.pdf`, October 2009.

[20] Ethernet Alliance, Energy Efficient Ethernet, `http://www.ethernetalliance.org/about_us/technology_overview/energy_efficient_ethernet`.

[21] Verizon Corporate Responsibility Report 2008–2009, `http://responsibility.verizon.com/images/vz_uploads/verizon_cr_report_2008-2009.pdf`.

[22] Sustainable energy use in mobile communications, Ericsson White Paper, `http://www.ericsson.com/campaign/sustainable_mobile_communications/downloads/sustainable_energy.pdf`, August 2007.

[23] S. Roy, Energy logic for telecom, White Paper from Experts in Business Critical Continuity, Global Marketing, Emerson Network Power, `http://www.emersonnetworkpower.com/energysystems/pdf/ES-113_EnergyLogic.pdf`, September 2008.

[24] D. Faulkner, Wireline networks, BT, ETSI Green Agenda, `http://portal.etsi.org/docbox/Workshop/2009/200911_GREENAGENDA/11Faulkner_Wireline%20Networks.pdf`, November 2009.

[25] Code of Conduct on Energy Consumption of Broadband Equipment, Version 3, European Commission Joint Research Centre, `http://re.jrc.ec.europa.eu/energyefficiency/pdf/CoC%20Brodband%20Equipment/Code%20of%20Conduct%20Broadband%20Equipment%20V3%20final.pdf`, November 2008.

[26] Universal charging solution explained, GSMA, `http://gsmworld.com/documents/Universal_Charging_Solution_Explained_v1.4.pdf`.

[27] NGNs and Energy Efficiency, ITU-T Technology Watch Report #7, August 2008.

[28] Energy Star* Version 5 Implementation, Intel White Paper, `http://download.intel.com/design/processor/applnots/321556.pdf`, February 2009.

[29] ITU-T Focus Group on ICTs and Climate Change – Report to TSAG, Direct and Indirect Impact of ITU Standards of Energy and Carbon Footprint, `http://www.itu.int/oth/T3307000006/en`, April 2009.

[30] Bureau of Energy Efficiency, Ministry of Power, Government of India, `http://www.bee-india.nic.in/`.

Biography

Prateep Misra has over 17 years experience in software development, quality assurance, R&D and project management. His areas of expertise includes process control and automation systems, digital signal processing, real-time & embedded systems design, RFID and sensors and IT infrastructure. Prateep has been instrumental in setting up the Center of Excellence of Digital Signal Processing and the RFID Technology Center in TCS. Prateep is an IEEE member and a certified software quality analyst. Prateep is a research area manager in the TCS Innovation Lab. He leads research programs in the area of open storage and complex event Processing. Dr. Prateep received his B.Tech in Instrumentation Engineering from the Indian Institute of Technology, Kharagpur in 1990 and his M.Tech in Control Systems Engineering from the same institute in 1993.

7

Conclusions

Ramjee Prasad

Founder Chairman, GISFI and Director CTIF, Aalborg University, Denmark

Standards are key to creating and ensuring interoperability and, hence, these contribute to avoid the fragmentation of markets. Therefore, the efforts of GISFI will harmonize the fragmentation in the market. This is of particular importance in rapidly evolving markets with ever changing technologies, notably in the ICT area.

The liberalization of the ICT market and other fundamental changes, such as the coexistence and convergence of heterogeneous ICT networks, that have occurred in the past years have led to more competition among players and sectors. On the back of such trends, ensuring interoperability becomes an even more critical issue.

The five ICT sectors described in this book and their trends and challenges are critical for the future of ICT. The common trend for each of these sectors is that there is a major shift from hardware to software that implies that standardization should focus more on short-term product development and exploitation in order to achieve interoperability among the offered technological solutions.

Thus, GISFI must work in close cooperation with the major ICT stakeholders in order to answer the demand of the changing market. In particular, in fast developing markets, powered by rapid ICT technological evolution, industry and GISFI efforts must be fully engaged with each other to achieve coherence of concepts and policies that answer the demands of the various types of markets. A reference to the importance of cooperation within standardization activities is the external evaluation of candidate technology

R. Prasad (ed.), Future Trends and Challenges for ICT Standardization, 271–275.
© 2010 *River Publishers. All rights reserved.*

proposals for IMT-Advanced adopted within the ITU-R process for the definition of the IMT-Advanced radio interface.

GISFI would seek to initiate a strategic review, together with various Indian and non-Indian stakeholders that would target how all players involved in standardization could better match the challenges responding to societal and market needs in India and worldwide, thus providing efficiently elaborated specifications in the ICT sector.

The motivation behind this book is that a more informed context is made available to ensure sustainable scientific and economic growth.

The future of ICT is characterized by a massive increase in the complexity and heterogeneity of the network infrastructure characterized by convergence, and support for sensors, mobility, and a variety of new highly dynamic services.

To handle this change it is necessary to design architectures and concepts for handling uncertainty and enabling an autonomous decision making process. This underpins a fundamental shift in the design focus, from performance-oriented design to design for robustness and resilience. New bottom-up paradigms (including bio-inspired approaches) will be required to address the issues of scalability, reliability, resilience, interoperability, security, and limitations of power, mobility, and spectrum.

While the data rate requirements of mobile applications are increasing and users are starting to use mobile devices more intensively, the available spectrum to provide the required capacity is severely limited. Therefore, it is important to make efficient use of the scarce spectrum resource. While it is continuously tried to further increase the spectral efficiency on the physical layer, a completely different approach is to make more flexible use of the spectrum. By getting away from the idea of a fixed and exclusive dedication of the spectrum it becomes possible to share spectrum in different ways to utilize it more efficiently.

Here, we have identified the current trends in spectrum technologies and management. The various existing standards answering the demands of the different world regions in the area of ICT actually consume further the available frequencies but there are still many frequency bands that are left unused. A common effort towards technologies allowing the reuse of spectrum and better spectrum management, as well as interoperability solutions can benefit this critical ICT area very much.

The success in the development of wireless communications has resulted in the installation of commercial cellular networks. The ease of use and importance of mobility has led to the huge increase in the use of cell phones.

In many cases such systems are limited in the number of mobile users they can simultaneously handle. Market competition drives the costs lower, which results in the increased number of users willing to pay for the wireless service. Unfortunately, bandwidth constraints and stringent regulations on the radiated power limit the capacity of such systems. Being limited in wireless resources, the only way to reduce costs and keep the attractiveness of wireless systems is to offer the highest possible efficiency of resource utilization, which directly translates into the efficient use of system capacity.

While cellular wireless systems are easily scalable they still have a number of unsolved problems. For example, one of them is related to the mobility management that assigns a mobile user to the certain location. Ad hoc type of connections, based on cooperation and sharing of knowledge are currently under research. In addition to that, a more flexible system based on user access priorities is also under consideration by regulation bodies.

Here, description of the current developments in the area of radio access technologies towards a future ICT is made. There is a significant effort and joint work form the various stakeholders towards ubiquitous radio systems answering the need for delivery of true broadband services anytime, anyhow, and anywhere. The current technological effort in this area relies very much on interoperability solutions, which implies the need for close cooperation among stakeholders and the various standardization bodies.

Convergence is an essential component of a future ICT infrastructure. Convergence ranges from rather simple convergence cases like fixed-mobile telephone services to much more complex cases involving digital content, networks, services and devices. To make the market profitable for all players, the current value chains and business models must be revisited and the main players must think of providing services in a more integrated way.

One convergence trend is the broadband and digital convergence, which is still considered the key driver for industry growth. Increasing bandwidth will enable and drive demand for an expanding range of services. Today, a platform for future converged communication and entertainment services is slowly emerging. This in turn makes a shift towards a growing demand for user individualization and the need to tailor products to suit the individual usage patterns, which is in essence the concept of on-demand entertainment services.

Further, it is explained how the horizontal integration of infrastructures, market and services could lead to strengthening of the market power. Convergence is also key to realizing true broadband services. Future wireless networks can be enhanced dramatically by taking a full advantage of the heterogeneous network environment, by applying spectrally more efficient transmission schemes, and by deploying novel network topologies. A key challenge is to coordinate and manage the different types of networks jointly. Therefore, interworking between these networks is essential. Interworking must happen in a generic and cooperative way to fully exploit the available resources and increase the performance gains of the systems.

IoT is a deep technological revolution that represents the future of computing and communication. The IoT scenario hosts the vision of ubiquitous computing and ambient intelligence enhanced by requiring a complete communication and computing capability among "things". The technologies of the Internet of Things offer immense potential to consumers, manufacturers and firms. Normally an ICT deeply influences the market trend providing business development and defining new economical path. IoT can be seen as an economical revolution. Such a scenario allows the pervasiveness of communication technologies in many economical and societal, with subsequent prospects of ICT based growth and wealth creation through innovation.

Here, we have identified the challenges in the world of IoT, where various devices, technologies and sectors are interlinked in a large-scale fashion. Interoperability is key for realizing real-world Internet and exploiting the economical and societal advantages offered by the IoT applications. Embedded systems and software, middleware platforms, sensor technologies are only some of the main enablers of the IoT. Security, privacy and reliability are other concerns that must be mapped to possible interoperability solutions. The development of IoT as enabler of the future ICT depends on the dynamic technical innovation in a number of important fields from wireless sensors to nanotechnology. A combination of all of these developments creates an IoT that connects the communicating objects in both a sensory and an intelligent manner.

ICT advances make it possible to detect environmental problems at very large and very small scales. ICT can help expand the use of renewable energies, reducing greenhouse gas emissions and increase energy efficiency. Critical areas where ICT applications have a significant effect for achieving environmentally sustainable economy are the energy sector, traffic management, reducing the waste and recycling, and so forth. We have identified the importance of interoperability for realizing environmental sustainability and making the ICT more energy-efficient. Each of the other thematic areas described in this book can contribute to the realization of a green ICT infrastructure. Therefore, a common platform of stakeholders and standards in such a "vertical" scenario must be realized.

The strategic framework realized by GISFI is focused on the technical research and development challenges (e.g., related to openness and connectivity, security and authentication, access and management of resources) and the goal is to propose strategies for their resolution that will create technical and development synergies. GISFI will provide equal awareness about research and development as well as policies and regulatory and standardization activities and strategies between the world regions, which would benefit the regulatory environment developments towards common goals.

About the Editor

Ramjee Prasad is currently the Director of Center for TeleInFrastruktur (CTiF), and holds the chair of wireless information and multimedia communications of Aalborg University, Denmark. Professor Prasad is also the Founding Chairman of GISFI. He has published over 700 technical papers, contributed to several patents, and has authored, co-authored, and edited over 30 books. His latest book is *My Personal Adaptive Global NET (MAGNET)*. Professor Prasad is the founding Editor-in-Chief of the Springer International Journal on *Wireless Personal Communications*. He was also the founding chairman of the European Center of Excellence in Telecommunications, known as HERMES and now he is the honorary chairman. He is a Fellow of IEEE, a Fellow of IET, a Fellow of IETE, a member of The Netherlands Electronics and Radio Society (NERG), and a member of IDA (Engineering Society in Denmark). Professor Prasad is advisor to several multinational companies. He has received several international awards, including the "Telenor Nordic 2005 Research Prize".

Index

RIVER PUBLISHERS SERIES IN STANDARDISATION

Volume 1
Wireless Independent Living for a Greying Population
Lara Srivastava
2009
ISBN: 978-87-92329-22-6

Volume 2
Entity Authentication and Personal Privacy in Future Cellular Systems
Geir M. Køien
2009
ISBN: 978-87-92329-32-5

Milton Keynes UK
Ingram Content Group UK Ltd.
UKHW031144141024
449569UK00024B/1088